Mineralogy - Significance and Applications

Edited by Ali Ismail Al-Juboury

Published in London, United Kingdom

IntechOpen

Supporting open minds since 2005

Mineralogy - Significance and Applications
http://dx.doi.org/10.5772/intechopen.83247
Edited by Ali Ismail Al-Juboury

Contributors
Violeta Nikolic, Anupam Misra, Tayro Acosta-Maeda, Ali Ismail Al-Juboury, Mohsin Ghazal, Sabhan Jalal, Miloš René, Zeynel Başıbüyük, Gökhan Ekincioğlu, Neha Sharma, Martin Onani, Leandré Bianca Brandt, Zuraan Paulsen, Jagadeesh Angadi V, I.C. Sathisha, K. Manjunatha, B. Chethan, Y.T. Ravikiran, Vinayaka K. Pattar, S.O. Manjunatha, Shidaling Matteppanavar, Nurit Taitel-Goldman, İlkay Kaydu Akbudak

First published in London, United Kingdom, 2020 by IntechOpen
IntechOpen is the global imprint of INTECHOPEN LIMITED, registered in England and Wales, registration number: 11086078, 7th floor, 10 Lower Thames Street, London, EC3R 6AF, United Kingdom
Printed in Croatia

British Library Cataloguing-in-Publication Data
A catalogue record for this book is available from the British Library

Additional hard and PDF copies can be obtained from orders@intechopen.com

Mineralogy - Significance and Applications
Edited by Ali Ismail Al-Juboury
p. cm.
Print ISBN 978-1-78985-825-9
Online ISBN 978-1-78985-826-6
eBook (PDF) ISBN 978-1-83880-738-2

For EU product safety concerns:
IN TECH d.o.o., Prolaz Marije Krucifikse Kozulić 3, 51000 Rijeka, Croatia, info@intechopen.com or visit our website at intechopen.com.

Meet the editor

Ali Al-Juboury is a professor in the Geology Department, University of Mosul, Iraq. He obtained his BSc in Geology and MSc in Sedimentology in 1980 and 1983, respectively, and his PhD from Comenius University, Slovakia, in 1992. He has published ninety-four papers in local and peer-reviewed journals in the fields of mineralogy, petroleum geology, sedimentology, and geochemistry. He is a member of numerous international societies and serves on the editorial board of the *Iraqi Geological Journal* and *International Sedimentology and Stratigraphy Journal of Oil and Gas Basins*, *Azerbaijan*, and *International Journal of Geophysics and Geochemistry* (American Association of Science and Technology). He has received several awards, such as the Distinguished Scholars Award from the Arab Fund for Economic and Social Development, Kuwait, in 2009, and the Science and Technology (Geology) Award from the Islamic States in 2014.

Contents

Preface

Mineralogy is a branch of geology that deals with chemistry, crystal structure and physical properties of minerals in various igneous, metamorphic and sedimentary processes, from high to low temperatures and in various surfaces to deep Earth's interior conditions and from continental to marine environments.

Mineralogy is also an important background and of high significance for many other scientific disciplines including environmental and geotechnical firms, mining and materials industry, oil and gas industry and biomedical science.

Minerals have a wide spectrum of applications and uses, such as industrial minerals, ores and gemology. Several society infrastructures are also made completely from mineral resources (e.g., cell phones, copper wires, diamond rings, gold jewelry, aluminum cans, steel bridges, plaster walls, nuclear reactors, and many others).

Mineralogy - Significance and Applications includes ten chapters that discuss the importance and applications of mineralogy in selected regions of the world. Generally, these contributions range geographically from various countries in Asia, Europe and Africa and extend far to Mars, with contributions on some industrial and biomedical applications of minerals.

Chapter 1, "Hematite spherules on Mars," by Anupam K. Misra and Tayro E. Acosta-Maeda focuses on the hematite spherules on Mars through the Mars Global Surveyor (MGS) launched by NASA and data gained from the NASA's Mars Exploration Rover "Opportunity," which landed in Eagle crater on Meridiani Planum. The authors discuss the observed properties of Martian hematite spherules and explain why a cosmic spherule formation mechanism provides a possible solution to the puzzling observations on Mars.

Chapter 2, "Mineral chemistry of Chalki basalts in northern Iraq and their petrological significance," by Mohsin M. Ghazal, Ali I. Al-Juboury and Sabhan M. Jalal describes the use of an electron probe microanalyzer (EPMA) to distinguish the various phases of minerals resulted from the alteration on basaltic rocks of the Chalki volcanics in extreme Northern Iraq, as well as interprets their petrologic significance.

Chapter 3, "Titanite from titanite-spots granodiorites of the Moldanubian Batholith (central European Variscan Belt)," by Miloš René integrates petrologic, mineralogic and geochemical analyses of titanite enclosed in hell "spots" in titanite-rich granodiorites from the Austrian Mühlviertel to show their significance in petrogenesis as late-magmatic evolution.

Concerning industrial applications of mineralogy, Chapter 4, "Investigation of the usability of pseudoleucites in Central Anatolia alkali syenites as industrial raw materials," by Zeynel Başibüyük and Gökhan Ekincioğlu studies intrusion-related distributions, mineralogical and petrographical properties, and mineral chemistry of pseudoleucites in İsahocalı (Kırşehir) alkali syenites from Central Anatolia granitoids. It also demonstrates that these pseudoleucites can be used as industrial raw material in many application such as ceramics, agriculture, and cement industries.

Chapter 5, "Mineralogical-petrographical investigation and usability as the gemstone of the North Anatolian kammererite, Tokat, Turkey," by İlkay Kaydu Akbudak, Zeynel Başibüyük and Gökhan Ekincioğlu focuses on nodules and thin veins of kammererites within Mesozoic basic-ultrabasic rocks from Turkey and their use in both jewelry and ornamental objects.

Chapter 6, "Enhanced humidity sensing response in Eu^{3+}-doped iron-rich $CuFe_2O_4$: A detailed study of structural, microstructural, sensing, and dielectric properties," by I.C. Sathisha, K. Manjunatha, V. Jagadeesha Angadi, B. Chethan, Y.T. Ravikiran, Vinayaka K. Pattar, S.O. Manjunatha and Shidaling Matteppanavar describes the synthesis of $CuFe_{(2-x)}Eu_xO_4$ (where x = 0.00, 0.01, 0.02, 0.03) nanoparticles via solution combustion using a mixture of fuels for the first time. Higher concentration and good-sensing ferrites are used for sensor applications, while low-sensing ferrites are used in battery and electronic applications.

Consequently, Chapter 7, "Iron oxides synthesized in hypersaline solutions," by Nurit Taitel-Goldman describes the synthesis of iron oxides using high-resolution scanning electron microscopy (HRSEM) in conditions similar to those prevailed in the Red Sea. The authors conclude that crystallized phases were submicron magnetite, feroxyhyte, goethite and akagenéite. The study reveals that iron oxides were synthesized at higher pH, elevated temperatures and hypersaline brines.

The significance and applications of mineralogy in medicine are addressed in the next three chapters.

Chapter 8, "Preparation and characterization of Fe_2O_3-SiO_2 nanocomposite for biomedical application," by Violeta N. Nikolić emphasizes the importance of investigating the influence of synthesis parameter variations onto the magnetic properties of composite materials containing nano-hematite particles. These materials could be used as starting materials for preparing multifunctional nanoparticles that can be used in different areas of biomedicine.

Chapter 9, "Future of nanoparticles in the field of medicine," by Neha Sharma concludes that magnetic nanoparticles can be used in hyperthermia treatment involving removal of tumorous cell/tissue with not much collateral damage.

Finally, Chapter 10, "Chemical synthesis and characterization of luminescent iron oxide nanoparticles and their biomedical applications," by Martin Onani, Leandre Brandt and Zuraan Paulsen discusses the synthesis of iron oxide magnetic nanoparticles using the co-precipitation method and their application as diagnostic and drug delivery tools for treatment of cancer and many other diseases.

This work was achieved with great support from fruitful suggestions of expert reviewers whose comments and contributions play a recognizable role in finalizing this book in high quality. We express our gratitude to Mohsin Ghazal, Angelo Paone, Igor Petrik, Hakan Çoban, Mohammad Aljaradin, Dicle Bal Akkoca, Marek Łodziński and Prof. Yasuhisa Maeda. We also acknowledge the scientific contribution by Dr. Brajesh Kumar, which has greatly benefited our book project.

Dr. Ali Ismail Al-Juboury
Professor,
Geology Department,
University of Mosul,
Mosul, Iraq

Chapter 1

Hematite Spherules on Mars

Anupam K. Misra and Tayro E. Acosta-Maeda

Abstract

In 2004, the observation of large amounts of hematite spherules on Mars by the NASA's Mars Exploration Rover "Opportunity," which landed in Eagle crater on Meridiani Planum, created tremendous excitement among the scientific community. The discovery of hematite was significant as it suggests past presence of water on Mars. Furthermore, the hematite spherules were widely suggested to be concretions that formed by precipitation of aqueous fluids. Among the various observed mysteries of Martian hematite spherules, also known as "blueberries," one regarding to their size limit was very puzzling. All of the millions of blueberries observed on Mars were smaller than 6.2 mm in diameter. Because the concretions on Earth are not limited in size, the formation of the Martian blueberries became difficult to explain. In this chapter, we will discuss the observed properties of Martian hematite spherules and explain why a cosmic spherule formation mechanism provides a possible solution to the puzzling observations on Mars.

Keywords: Martian hematite spherules, Martian blueberries, cosmic spherules, concretions

1. Introduction

In 1996, NASA launched the Mars Global Surveyor (MGS) spacecraft to perform global mapping of Mars. One of the instruments on the MGS is the thermal emission spectrometer (TES), which would map the mineralogy of the Martian surface using infrared spectroscopy. TES imaging revealed the presence of crystalline gray hematite on Mars in Sinus Meridiani. **Figure 1** shows the global distribution of minerals on Mars [1, 2]. The distribution of hematite is shown in pink and labeled as H in the areas of Aram Chaos and Sinus Meridiani. The bottom image shows the distribution of hematite in Sinus Meridiani. According to [1] and [3], the hematite covers an area of over 175,000 km^2. The hematite boundary is abrupt and immobile and all the hematite is very possibly confined to a thin layer. According to [3], this layer could be only 100 microns, because TES gives surface measurements. The age of hematite is estimated over 3.5 Ga. The unnamed crater shown in the bottom image (**Figure 1**) shows no hematite, indicating that it was formed after the hematite deposit. Similarly, the inflow from top may be newer than hematite. The authors suggested that these hematite deposits were formed by chemical precipitation from aqueous fluids, and TES data provide evidence that liquid water has been stable for millions of years on early Mars.

The Mars Exploration Rover "Opportunity" landed in Eagle crater on Meridiani Planum in the western part of the Sinus Meridiani region on January 24, 2004. Within a few days of landing on Mars, the Opportunity rover sent pictures of large numbers of spherules, as shown in false-colored images in **Figures 2** and **3** [4].

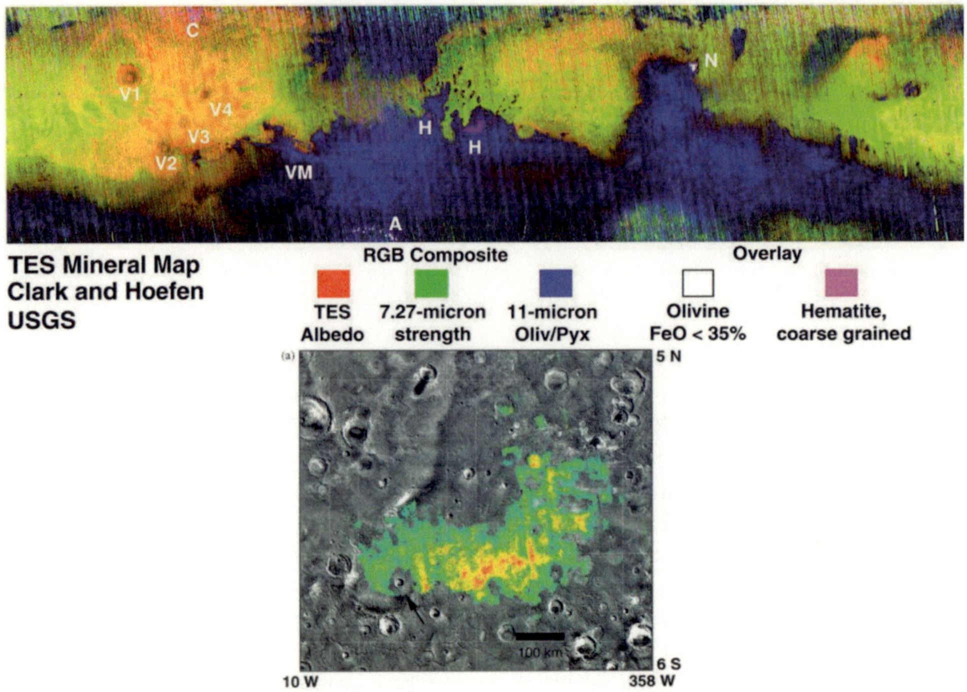

Figure 1.
Global distribution of hematite on Mars using the thermal emission spectrometer on the Mars Global Surveyor Spacecraft (top image). Bottom image is a close-up view of hematite abundance in Sinus Meridiani. (Image courtesy of USGS, NASA/JPL).

Figure 2.
Observations of hematite spherules at Meridiani, Mars, by the Opportunity rover. (Image courtesy of NASA/JPL).

The spherules were studied using instruments on board the Opportunity rover. The Mössbauer spectrometer was used to confirm the mineralogy of the spherules as hematite. A rock abrasion tool (RAT) was used to cut some of the spherules and concluded that spherules are also very hard. The rover instruments provided ground validation data confirming the presence of hematite on Mars as predicted by the orbital TES data obtained by the MGS spacecraft. The gray hematite spherules

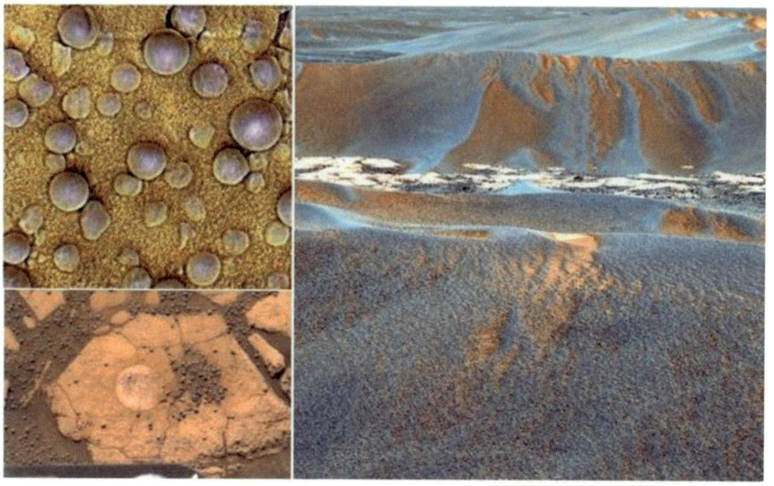

Figure 3.
Observations of hematite spherules at Meridiani, Mars, by the Opportunity rover. (Image courtesy of NASA/JPL).

appeared blue in the false-colored data from Mars and were therefore nicknamed "blueberries." The discovery of Martian blueberries quickly became an exciting scientific discovery as leading scientists concluded that hematite spherules were concretions and that their discovery proved the presence of water in Mars history.

The idea that the Martian hematite spherules are concretions has been largely accepted by most planetary scientists for the past several years [3, 5–15]. Several scientists also found concretion terrestrial analogues in southern Utah, in the Jurassic Navajo Sandstone [16], and in Lake Brown, Australia [17]. In the next section, we will discuss the observed properties of the Martian hematite spherules, which will lead us to the controversy of the origin of these spherules.

2. Observed properties of Martian hematite spherules

Within a few days of operations, the Opportunity rover surprised the science community by sending pictures of a large number of spherules on Mars, now commonly known as blueberries [18]. The follow-up investigation [11, 13] by the NASA science team found several interesting observations of blueberries. Some of these observations are as follows: (i) the primary carriers of hematite at the landing site are blueberries and their fragments; (ii) the hematite was located mostly on the surface of the landscape; (iii) the deeper soil is mostly basaltic sand and is free of hematite; (iv) the blueberries are mostly perfect spheres; (v) the typical diameters of the blueberries are 4–6 mm; (vi) the blueberries are hard; (vii) the blueberries are made of very fine grain material; and (vii) the blueberries have no internal structure. In addition, all the hematite spherules appear to be located within the upper 10 mm thickness of the surface soil.

Figure 4 shows the result of the RAT used on the Martian surface to cut open some of the soil-embedded spherules. The spherules were found to be hard and difficult to cut. The spherules showed no internal structure with grain size below the detection limit (31 μm) of the Microscopic Imager (MI). Further investigation of the spherules found that there are two types of blueberries: larger blueberries with average size of 4 mm in diameter and microberries with average size of 0.79 mm in diameter. All blueberries were smaller than 6.2 mm in diameter with a median size of 4.2 mm.

Figure 4.
Rock abrasion tool (RAT) was used to cut some of the spherules embedded in the soil. The hematite spherules show no internal structure. (Image courtesy of NASA/JPL).

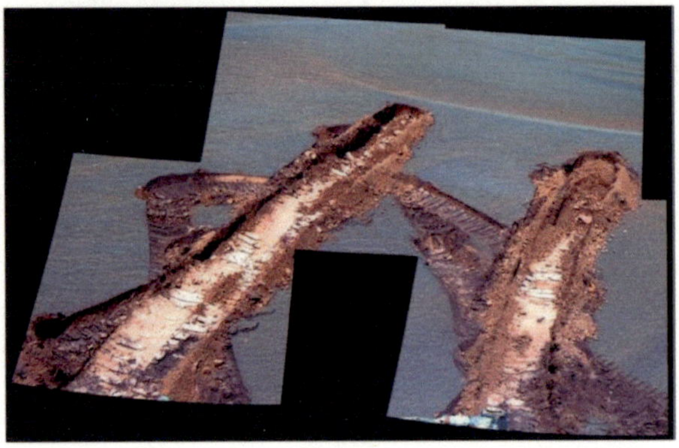

Figure 5.
Jammerbugt (sol 842) showing the trenches dug by the Opportunity rover. All hematite is located on the Martian surface and no blueberry was found in the deeper soil. (Image courtesy of NASA/JPL).

On sol 833, Opportunity got stuck in a fine-grained soil (named Jammerbugt) and it took 8 days for the operation team to free the rover. During this process, the rover's wheel dug deep trenches on Martian soil. **Figure 5** shows images of Jammerbugt taken by the rover on sol 842 (June 7, 2006) using the panoramic camera. **Figure 5** shows that all the hematite (blueberries, microberries, and fine dust) is limited to the top surface and trenches dug by the rover showed no sign of blueberries in deeper soil.

In summary, a very large amount of Martian hematite spherules was found to be mostly perfect hard spheres less than 6 mm in diameter with fine grain, no internal structure, and located within 10 mm of the Martian surface.

3. Comparison of Martian hematite spherules and Earth's concretion analogues

It has been suggested by various scientists that the Martian hematite spherules are concretions. Because concretions are formed by water on Earth, this carries the significant scientific implication of proof of water in Martian history. We will first examine the properties of Earth's concretions to see if they are suitable for classifying them as terrestrial analogues of Martian spherules. A concretion is a compacted mineral body that is embedded in a host rock, which has a different chemical composition. Terrestrial concretions are formed from liquid phase by precipitation, nucleation, and growth processes.

Figure 6.
Earth's concretion analogue samples from the Jurassic Navajo Sandstone, southern Utah (courtesy William Mahaney). A 6.4 mm diameter steel ball is also shown for size reference.

The concretions can grow in size and are found in various sizes and shapes. They are rarely perfect spheres and have no size limitation. Examples of terrestrial concretions (Jurassic Navajo Sandstone, southern Utah), which have been cut to show their internal structure, are displayed in **Figure 6**. The concretions are usually round objects with heterogeneous internal structure and are rarely perfect spheres. They are not limited in size and can grow to a diameter of several centimeters and even meters. **Figure 6** also shows a steel ball with a diameter of 6.4 mm. The blueberries on Mars are all less than 6.4 mm in diameter and have similar physical properties to the steel ball.

Next, we look at the chemical composition of the concretions found on Earth. Raman spectroscopy is considered as a fingerprint technology for chemical identification. Raman spectra represent the vibrational modes of a molecule and all different chemicals have unique Raman spectra; for example, no two chemicals have the same Raman spectrum. Hence, Raman spectra can identify a chemical with 100% confidence level. Presently available commercial micro-Raman systems are capable of identifying small mineral grains in the nanogram range. Micro-Raman spectroscopy with 785 nm laser excitation has been used to identify the chemical compositions of Earth's concretions. Raman spectra were measured at multiple locations on all the samples. **Figure 7** shows representative Raman spectra of hematite, rose quartz, and goethite with both brown and black grains of Earth's concretions. The Raman spectra of the brown grains of concretions, which are located inside of concretions, are same as the Raman spectrum of quartz. This confirms that the interior composition of Earth's concretions is quartz with grain size of the order of 150 microns. The outer darker layers of the concretions also show the quartz Raman band along with Raman peaks of goethite. This suggests that the dark layers are made of quartz coated with a thin layer of goethite. This chemical analysis of Earth's concretions suggests that they are not hematite, which is consistent with reference [19].

In the formation of concretions, water containing dissolved minerals cement around the grains of the host matrix, which is why grains of host matrix are always included in the Earth's concretions [17]. In fact, none of the concretions on Earth are made of pure hematite; they can be hematite-coated quartz or calcite. The concretion mechanism cannot remove the grains of the host matrix (quartz, calcite, etc.) and replace them with pure hematite mineral. It may be possible to form pure

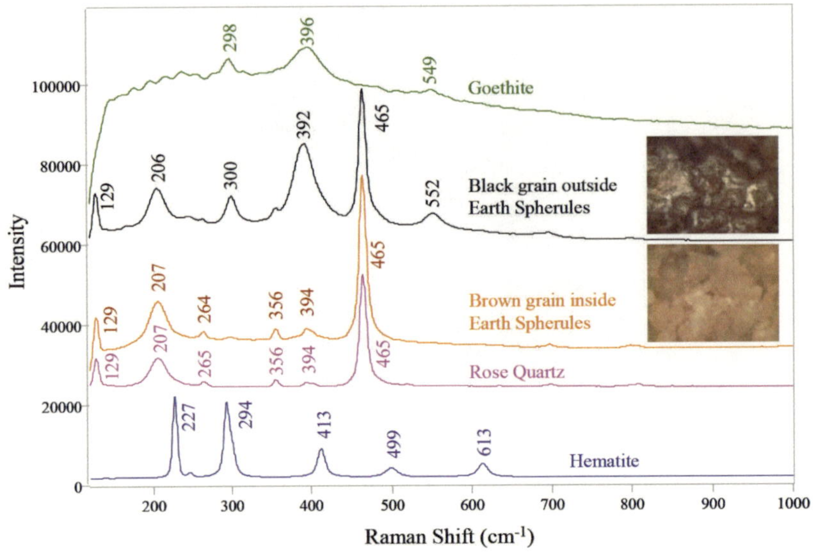

Figure 7.
*The Raman spectra of concretions shown in **Figure 6** confirm that the interior of the concretion is mainly quartz. The outer dark grains are made of quartz, which are coated with goethite.*

hematite crystals from an aqueous solution but it will have the shape of a crystal and not a perfect sphere. A recent attempt to form spherules by freezing an aqueous hematite nanoparticle suspension failed in a laboratory setting [20]. The concretion model does not explain the following: (i) why are the Martian hematite spherules limited in size? (ii) why are the spherules pure hematite? and (iii) why are grains of the host soil missing from the interior of the spherules? The formation of concretion from aqueous media also leads to another interesting fact: concretions are formed deeper in the soil. This is because during the dry season, the level of ground-water goes down. This increases the concentration of dissolved chemicals in the subsurface soil, which favors the formation of concretions in deeper soil. For Earth's concretions, a relatively larger number are observed in the deeper soil [17] than the top surface layer. In contrast, the Martian blueberries are mostly concentrated within 1 cm of the top surface [11, 13, 17, 21]. No blueberries were observed in the deeper soil when trenches were excavated on Mars [6, 11–13] as shown in **Figure 5**.

In addition, Earth's concretions are not as shiny as some of the Martian blueberries. This is because on Earth erosion plays a critical role in the formation of concretions. Concretions are formed inside the host matrix and are released from the host matrix by eroding away the surrounding material. The erosion of surrounding material takes several thousand years. Therefore, it is easy to see signs of erosion, pitting, and flow patterns due to the presence of aqueous media on Earth's concretions, which appear as dull metallic-looking objects [16, 17]. The erosion process plays a critical role in the formation of concretions and dictates that the Earth's concretions are very old [17].

Figure 8 shows an image (sol 251) of a large 1 m long rock, "Wopmay," found on the Endurance Crater, showing hundreds of blueberries on its surface. The NASA science team suggested that the blueberries in the region are embedded in the rocks and eroding from them [11, 13]. One of the problems with this conclusion is that all the blueberries observed on the rock are fully exposed spherules and no blueberry is seen emerging from the rock. It is expected that the erosion process would reveal some partially exposed blueberries on the rock if they are concretions. In addition, rock erosion produces soil that has a similar reddish color as the rock and would

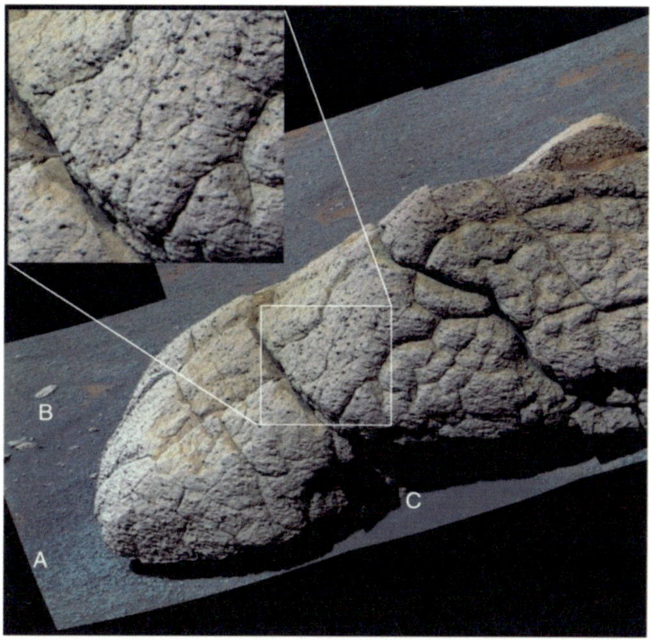

Figure 8.
Image showing hundreds of fully exposed blueberries on "Wopmay Rock" and nearby area. No blueberries are observed in region C (sol 251). (Image courtesy of NASA/JPL).

be mixed in with the bluish-looking soil. Lastly, the distribution of blueberries in the region and around the rock is consistent with the hypothesis that they fell from above. Region A, which is near the slope of the rock, shows a high concentration of blueberries as compared to region B, which is away from the rock. Also, in region C, in the shadow of the rock, there are no spherules, which contradicts the concretion mechanism. Another interesting observation is that spherules are heavy and therefore not easily transported by dust devil events. Otherwise, there would be no spherules on the slopes of the rock and region C would be filled with spherules. The observation of spherules on Wopmay rock suggests that blueberries fell from the top and are not eroding out of the rocks. The hypothesis of deposition of blueberries from the top is also consistent with the observation that the entire blueberry inventory is within 1 cm of the top surface layer.

4. Blueberries as cosmic spherules

There are two possible methods for depositing large number of hematite spherules from the top: (1) meteorite deposition and (2) volcanic deposition. Out of these two models, we suggest that the meteorite model is more consistent with all the observations of blueberries on Mars because, at present, there are no active volcanoes on Mars. Later, in this chapter, we will see evidence that some of the blueberries are very young as they have recently landed on the rovers and heat shield. This also favors the meteorite theory over volcanic deposition. According to the meteorite theory, meteorites of various sizes enter the Mars atmosphere at high speed and low temperature. When meteorites enter the Martian atmosphere, they feel friction and ram pressure, which heats up the meteorite. On Earth, commonly observed shooting stars suggest that heating can achieve very high temperatures, which make the meteorites glow. Under Martian conditions, the smaller meteorites can be completely melted. The liquid melts and then breaks down immediately into

Figure 9.
Image of "Heat Shield Rock," an iron meteorite observed on Mars. Several blueberries and microberries are observed in the close vicinity of the meteorite. Image taken from Sol 352, courtesy of NASA/JPL/Cornell.

smaller spherical drops whose size is determined by the surface tension of the liquid and the atmospheric drag force. Smaller drops soon achieve terminal velocity due to lower mass, which causes the temperature of the liquid drop to fall and become solid. Depending on the size of the meteorite and the time of flight, some of the drops will hit as solid balls and others as liquid drops that form microberries on collision with the ground [22]. Larger meteorites need more time to melt completely because of their mass. In bigger meteorites, melting begins at the meteorite surface, and as soon as the surface liquid reaches a critical depth, the liquid falls away as drops. This prediction suggests that the fusion crust on a meteorite should be also limited in thickness, and meteorites should also show imprints on the surface matching the size of the liquid drop, which was removed from the surface.

On sol 339, the Opportunity rover observed an iron meteorite on Mars, which was named "Heat Shield Rock." **Figure 9** shows an image of Heat Shield Rock, which is mainly iron with kamacite as the primary Fe-bearing mineral and around 7 wt.% nickel [23, 24]. A first look at the immediate surroundings of the meteorite confirms the presence of large number of microberries and blueberries near the meteorite. The surface of the fusion crust shows regmaglypts and several circular imprints. The circular imprints give an estimate of the size of the molten drops that fell off from the meteorite. This size matches with the size of blueberries lying on the ground. Several microberries and miniberries also form on meteorite impact with the ground from the liquid layer, which is still attached to the meteorite before impact. These microberries and miniberries will be distributed near the meteorite. A larger concentration of smaller blueberries and nano–dust particles forming a halo can be seen in the immediate surroundings of the meteorite (**Figure 9**). The image also shows several spherules on top of the meteorite. This image provides very strong direct evidence that Martian blueberries are cosmic spherules formed from the ablation of a meteorite. In addition, the same image also provides a strong evidence that blueberries are not concretions as iron meteorites do not carry enough water for concretion formation.

The formation of blueberries through molten meteorite drops also explains why all the observed blueberries on Mars are limited in size and are mostly perfect spheres with no internal structure and fine grain size. The size and shape of the blueberries are determined by the phenomenon commonly known as surface tension. On Earth, we can use raindrops as analogues: their sizes and shapes are controlled by the surface tension of water. The size of a raindrop can be estimated by assuming that the water drop will break up if the atmospheric drag force is greater than the surface tension force. When the gravitational force equals the atmospheric drag force, the raindrops achieve terminal velocities. These conditions give the estimated diameter of the raindrop (D) to be proportional to [25]:

$$D \propto \sqrt{\frac{\sigma}{g\rho}} \tag{1}$$

where σ is the surface tension of the liquid, g is the acceleration due to gravity, and ρ is the density of the liquid. The above equation can be simplified using two different types of liquids to give equation:

$$\frac{D_2}{D_1} = \sqrt{\frac{\sigma_2 g_1 \rho_1}{\sigma_1 g_2 \rho_2}} \tag{2}$$

where subscripts 1 and 2 represent two different types of liquids. By assuming that the melting of iron meteorites forms spherules, we can estimate the diameter (D_2) of Martian spherules. Using σ_1 = 0.073 N/m surface tension of water, g_1 = 9.8 m/s^2 as Earth gravity, ρ_1 = 1000 kg/m^3 as density of water, σ_2 = 1.46 N/m surface tension of molten iron [26], g_2 = 3.675 m/s^2 as Mars gravity, and ρ_2 = 7860 kg/m^3 as the density of iron gives D_2 = 2.6 D_1. According to the U.S. Geological Survey (USGS) website, raindrops that are spherical in shape are usually 1–2 mm in size (**Figure 10**). Larger drops with diameter of 3 mm are sometimes formed but are not spherical in shape [27, 28]. When the raindrop reaches a diameter of 4.5 mm, it opens up like a parachute and immediately breaks up into smaller drops and microdrops. One of the important features of the meteorite model is that it puts a size limit on the diameter of hematite spherules on Mars. Accordingly, hematite spherules on Mars would be spherical in shape if the diameter is less than 5.2 mm. Larger spherules are not expected to be perfect spheres and no spherules are expected to reach the size of 12 mm, which corresponds to immediately breaking up into smaller spherules and microspherules by opening up like a parachute. The meteorite theory correctly predicts the size limit of millimeter-size blueberries and also predicts the formation of a large number of perfectly spherical microberries and nanophase material.

One of the puzzling observations on Mars is that there are very large numbers of blueberries (**Figure 3**) and the vast majority of spherules are isolated spheres. The Opportunity rover observed doublets and triplets [3, 5, 13, 15] as shown in **Figure 11** but these instances were very rare. It is important to note that the observation of large numbers of blueberries and microberries does not suggest a large number of meteoritic

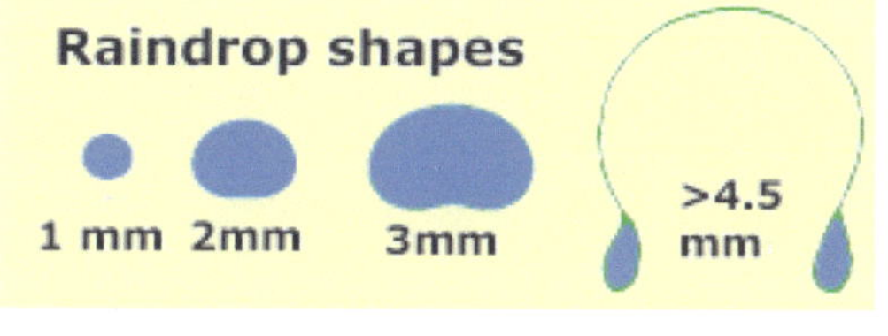

Figure 10.
Raindrop shapes with various sizes. (Courtesy USGS website).

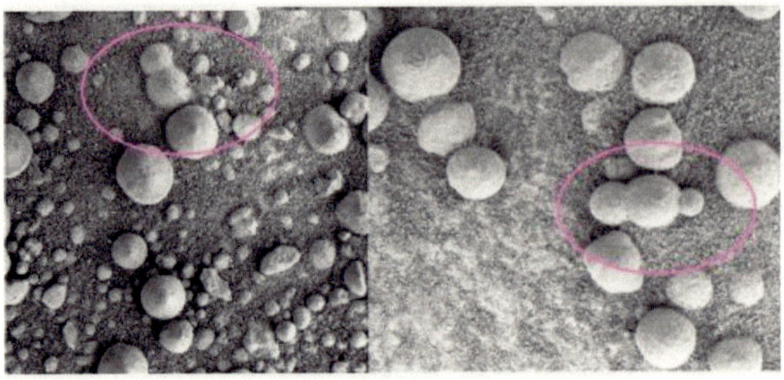

Figure 11.
Observed doublet and triplet blueberries on Mars.

events on Mars. A single small meteorite can produce a large number of spherules. For example, a 2-inch diameter meteorite is equivalent to 2048 spherules with diameter of 4 mm. In addition, a small meteorite entering the Martian atmosphere will distribute thousands of spherules over a large area on Mars along its trajectory. A very large number of spherules could also be formed during a meteorite shower event. Another important observation is that the population of doublets and triplets is very low in comparison to isolated spherules, which can be explained by the meteorite ablation mechanisms. The doublets can be formed when two liquid drops come in contact. However, in the liquid phase, this simply forms a bigger drop and immediately splits up into smaller drops. Similarly, two solid spherules will simply stay as two individual spherules and not form a doublet. For a doublet to form, the recombination of two spherules must occur near the liquid-solid phase transition. This significantly reduces the probability of the formation of a doublet or triplet. The meteorite model also predicts that doublets and triplets are more likely to have spherules of different diameters than the same diameter. This is because the doublet is more likely to form when one drop is moving faster than the other drop. Because the terminal speeds of drops are proportional to their masses, the doublets and triplets would be composed of spherules of different sizes as shown in **Figure 11**.

Because meteorites can fall on Mars at any time, there are other requirements that need to be satisfied for a meteorite ablation model. The first requirement is that new young cosmic spheres should be observed along with older spherules. The second is that cosmic spherules of other compositions should also be observed. **Figure 12** shows very shiny younger spherules among the older spherules.

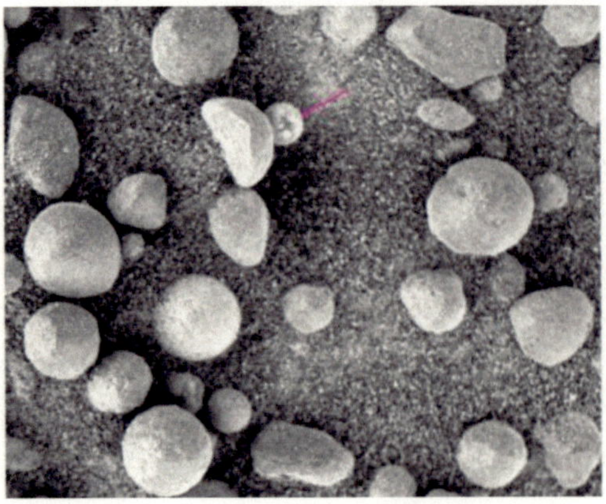

Figure 12.
Coexistence of younger and older spherules on Mars.

Figure 13.
Coexistence of yellow berries along with blueberries.

This observation is in agreement with the meteorite ablation model. However, it is very difficult to explain the coexistence of old and new spherules on Mars with the concretion model, which will require evidence of water appearing, disappearing, and reappearing on Mars. **Figure 13** from sol 319 shows the coexistence of both blue and yellow berries. In this false color image, the color indicates a change in optical reflectance, which indicates different chemical compositions for the blue and yellow berries. In the next section, we will look at the evidence that strongly supports the cosmic spherule mechanism.

5. Further evidence of cosmic spherules

The observations on Mars made by rovers shown in this section can only be explained by a meteorite model. There is a very small probability that man-made objects such as the rovers and heat shield will collect cosmic spherules from recent meteoritic events. However, for the meteorite model to be correct, there are few conditions: (1) the sizes of the spherules must obey the predicted size limit and (2) the spherules must look shiny in comparison to older blueberries, indicating younger age because the rovers landed on Mars in 2004.

Figure 14 from sol 339 shows a piece of the heat shield of the Opportunity rover. Because blueberries are spherical, they are most likely to collect at the bottom of a slope on the heat shield. The photograph clearly shows several fresh-looking blueberries collected on the heat shield. The size and texture can be compared with the older blueberries seen on the ground.

The images taken by the Opportunity rover of itself are shown in **Figure 15** (top image: sol 322; bottom image: sol 323), which show few blueberries collected by the rover. The size of these shiny spherules are similar to the size of the older, dull-looking blueberries lying on the ground as seen in the top right corner of the image shown in **Figure 15**. Further strong evidence in support of meteorite model is shown in **Figure 16**. If the spherules are formed by molten drops of falling meteorites, then they are expected to be very hot. **Figure 16** shows the image (1M156679326EFF3981P2979M2M1) taken on sol 321 of the solar panel by the microscopic imager on Opportunity. The image shows a burned impact mark at location 'B' along with few microspherules nearby shown as 'A.' The burn mark size and shape match the nearby spherules.

The other Mars rover "Spirit" landed in the Gusev Crater area. According to the MGS satellite image, this site is not rich in hematite. Hence, Spirit rarely observed spherules on the ground. The images from the Spirit rover taken on sol 330 show a few spherical objects on the solar panels marked by green circles (**Figure 17**, left).

Figure 14.
Evidence of fresh blueberries collected by heat shield. Image 1P158281536EFF40C2P2368L6M1 (sol 339). Image courtesy: NASA/JPL & M Lyle.

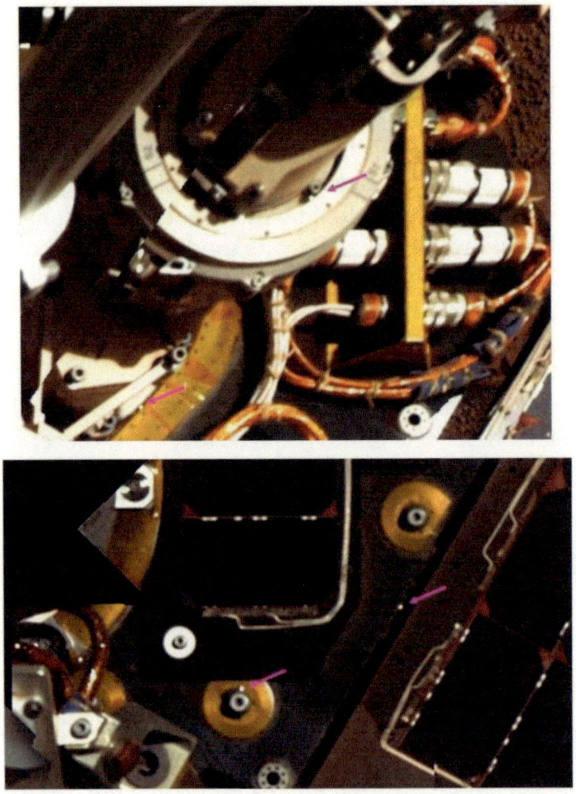

Figure 15.
Images of Opportunity rover showing shiny spherules collected by the rover. Sol 322 (top image) and Sol 323 (bottom image).

An image (**Figure 17**, right) of the same solar panels taken on sol 583 almost a year later shows that two of the objects have rolled off leaving behind the impact prints. A few new spherical objects have appeared near the electrical wirings. The ability

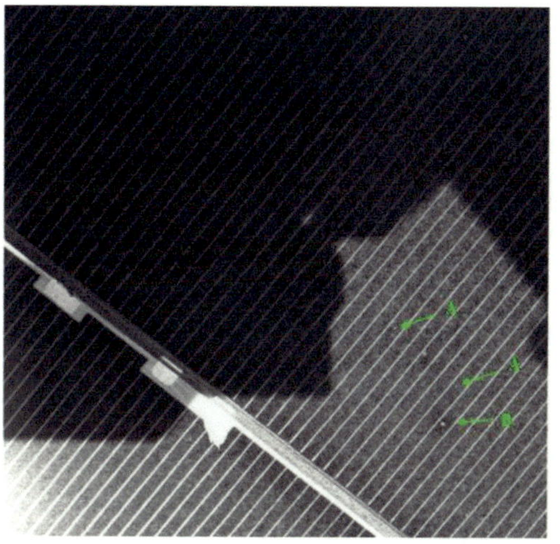

Figure 16.
Opportunity solar panel sol 321.

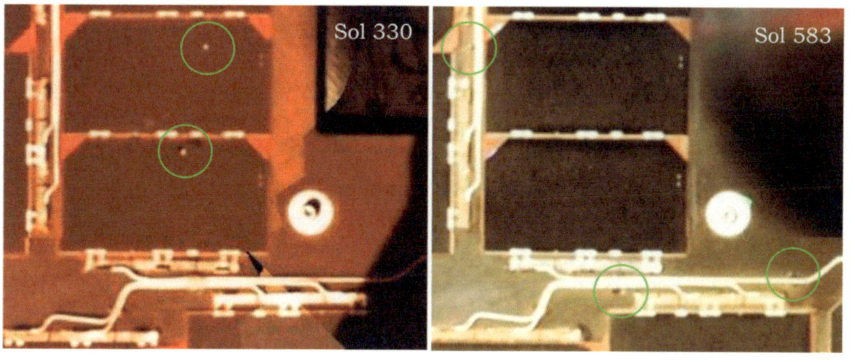

Figure 17.
Images comparing the same solar panels of Spirit rover from sol 330 and sol 583 showing blueberries that have rolled off a year later from solar panels.

of these round objects to leave impact marks on the solar panels indicates the high terminal speed of the spherules.

6. Discussion

The various observations of Martian blueberries strongly support the theory that the spherules are cosmic spherules formed from the ablation of meteorites. The images shown in **Figures 9** and **14–18** can only be explained by the meteorite model. In the meteorite ablation model, the cosmic spherules are formed from the liquid phase. Hence, they will be perfect spheres, hard, and size limited. The interior of the spherules will be glassy or extremely fine grained, and will not show nucleation or inclusions of basaltic grains of Martian surface. The blueberries, microberries, and nanophase material will be located only on the top soil and missing from the deeper soil. To date, all observations (listed in **Table 1**) made on Mars are consistent with the meteorite ablation model. To the best of our knowledge, no hematite concretions have been found on Mars and concretions made of pure hematite do not exist on Earth. The cosmic spherule mechanism also suggests that hematite found on the surface of Mars is extra-Martian and not native to Mars.

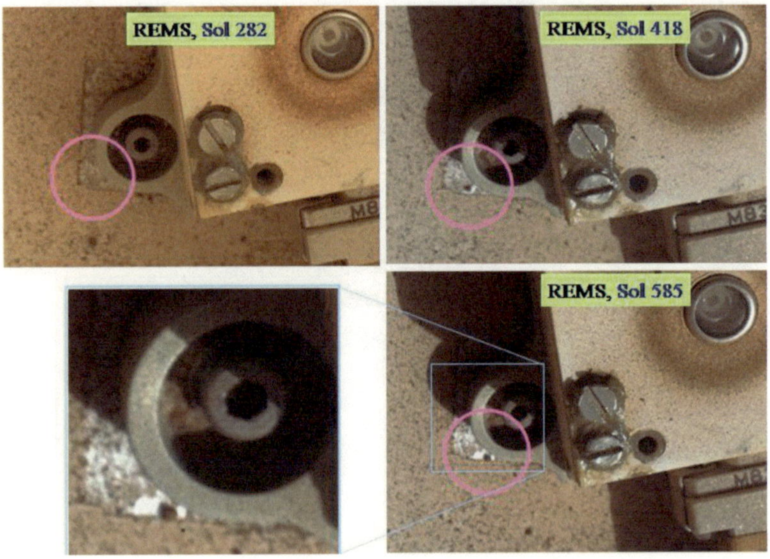

Figure 18.
Images of REMS-UV sensor on Curiosity rover showing a cosmic spherule impact.

A big iron meteorite or a few meteor shower events can produce a large number of spherules on Mars. About a billion spherules of 4 mm in diameter can be produced by a 4 m diameter meteorite. Because the spherules are heavy, the distribution of spherules at the Meridiani is expected to be sharply defined, immobile, and elliptical due to trajectories of the meteors. The age of the hematite deposit on Meridiani has been suggested by reference [3] to be in excess of 3.5 Ga. Opportunity rover did discover several large iron meteorites in the Meridiani Planum [23, 29]. Some scientists [30] have argued that the six iron meteorites found at Meridiani are the result of a single rare event of a large meteorite impact.

Mössbauer data from the Opportunity rover concluded that blueberries at Meridiani Planum are made of hematite [18]. It is possible to form hematite in a CO_2 atmosphere at temperatures above 900°C from high-FeO glass-rich basalts [31]. One possible redox reaction for the formation of hematite from iron meteorite is $2Fe + 3CO_2 = Fe_2O_3 + 3CO$. For meteorites that have some iron, it is possible to form hematite-coated spherules. Cooper et al. in Ref. [32] determined that at 700°C and FeO concentration > 1.9 wt%, the Fe migration is favored over Ca and Mg migration. The evidence for the formation of hematite particles at 60 km altitude on Earth has been suggested by [33].

It is interesting to note that Earth's Moon has no atmosphere and meteorites on the Moon do not get heated due to drag force. The spherules formed on Moon are mostly due to impact heating and are known as impact glass spherules. On Earth, millimeter-sized cosmic spherules are found in abundance [34–37]. The large amount of cosmic spherules on Mars could be due to its proximity to the asteroid belt [23]. Mars also has low gravity and thin atmosphere that are favorable conditions for forming large spherules.

A direct evidence of a cosmic spherule on Mars is shown in **Figure 18**. In 2012, NASA's "Curiosity" rover landed in Gale Crater. **Figure 18** shows images of the REMS-UV sensor, which is placed within the rover deck facing the sky. The sol 418 image shows that a small cosmic spherule has landed on the UV sensor, which was not observed in a previous image (Sol 282). The image taken on sol 585 shows that the spherule has rolled slightly to the right revealing the original impact spot.

Observations	Agreement with meteorite model
1. Millions of blueberries found on Mars (population)	Yes
2. All blueberries are less than 6.2 mm in diameter (size limitation)	Yes
3. Predominantly perfect spheres (shape)	Yes
4. Blueberries show no grain structure (internal structure)	Yes
5. Blueberries show no nucleation (mechanism)	Yes
6. Coexistence of old and fresh blueberries (age difference)	Yes
7. Coexistence of blue and yellow berries (chemical difference)	Yes
8. All blueberries limited to top soil and missing in deeper soil (location)	Yes
9. Interior of blueberries missing grains of host soil (pure phases)	Yes
10. Hematite (composition)	Yes[†]
11. Large amount of microberries observed only on top surface (population and location)	Yes
12. Blueberries appear embedded in soil (location)	Yes
13. Hard (mechanical strength)	Yes
14. Circular burn spot on solar panel (hot impact)	Yes
15. Rare observation of doublets and triplets (mechanism)	Yes
16. Doublets and triplets are made of different-sized spherules (mechanism)	Yes
17. All blueberries on Wopmay rock are fully exposed (mechanism)	Yes
18. Blueberries found on man-made objects (heat shield and Opportunity and Spirit rovers) (recent events)	Yes
19. Blueberries on heat shield and rovers are shiny (young age)	Yes
20. Blueberries found on and near iron meteorites (mechanism)	Yes

[†]*Other chemical phases are also possible.*

Table 1.
Observations and physical properties of Martian blueberries and their comparison with a meteorite model.

7. Conclusion

Some of the important observations of Martian blueberries cannot be explained by a concretion model. These observations include the following: (1) spherules are size limited, (2) they are located only on the top soil, (3) they show no internal structure, and (4) they lack grains of the host matrix. In addition, the distribution of spherules suggests that they fell from above, as shown in **Figure 8**. The observations of spherules collected by the heat shield and rovers suggest that these spherules are very young and cannot be explained by the process of aqueous alteration, which requires a significantly longer period of time. The observations of spherules and nanophase materials near the meteorites are direct evidence that spherules are meteoritic in nature. The meteorite ablation model producing cosmic spherules on Mars explains all the observations and properties of Martian blueberries. According to this mechanism, while traveling through Martian atmosphere, a meteorite gets very hot, reaches melting temperatures, and forms liquid molten drops that reach terminal speeds due to drag force and cool down to solid spherules and microspherules. The maximum size of the spherules is limited by the surface tension of the molten material and atmospheric drag force. The spherules are expected to be mostly perfect hard isolated spheres, with no internal structure and nucleation, and located only on the top surface layer. The meteorite mechanism also suggests that hematite found on the surface of Mars is extra-Martian.

Acknowledgements

All the data and images from Mars are courtesy of NASA and JPL. JPL image policy allows the use of images without prior permission as cited on their website http://www.jpl.nasa.gov/images/policy/

Author details

Anupam K. Misra* and Tayro E. Acosta-Maeda
Hawai'i Institute of Geophysics and Planetology, University of Hawai'i, Honolulu, Hawai'i, USA

*Address all correspondence to: anupam@hawaii.edu

References

[1] Christensen PR, Morris RV, Lane MD, Bandfield JL, Malin MC. Global mapping of Martian hematite mineral deposits: Remnants of water-driven processes on early Mars. Journal of Geophysical Research: Planets. 2001;**106**(E10):23873-23885

[2] Clark RN, Hoefen T. TES Mineral Map. USGS Website. Available from: https://speclab.cr.usgs.gov/mars.press.release.10.2000.html

[3] McLennan SM, Bell JF, Calvin WM, Christensen PR, Clark BC, de Souza PA, et al. Provenance and diagenesis of the evaporite-bearing burns formation, Meridiani Planum, Mars. Earth and Planetary Science Letters. 2005;**240**(1):95-121

[4] NASA/JPL. MER images from Mars. Available from: https://photojournal.jpl.nasa.gov. https://mars.nasa.gov/mer/gallery/images.html. http://mars.lyleresearch.com/

[5] Calvin WM, Shoffner JD, Johnson JR, Knoll AH, Pocock JM, Squyres SW, et al. Hematite spherules at Meridiani: Results from MI, mini-TES, and pancam. Journal of Geophysical Research: Planets. 2008;**113**(E12S37):1-27

[6] Christensen PR, Wyatt MB, Glotch TD, Rogers AD, Anwar S, Arvidson RE, et al. Mineralogy at Meridiani Planum from the mini-TES experiment on the Opportunity rover. Science. 2004;**306**(5702):1733

[7] Glotch TD, Christensen PR, Sharp TG. Fresnel modeling of hematite crystal surfaces and application to martian hematite spherules. Icarus. 2006;**181**(2):408-418

[8] Glotch TD, Bandfield JL. Determination and interpretation of surface and atmospheric miniature thermal emission spectrometer spectral end-members at the Meridiani Planum landing site. Journal of Geophysical Research: Planets. 2006;**111**(E12S06):1-12

[9] Kula J, Baldwin SL. On hematite as a target for dating aqueous conditions on Mars. Planetary and Space Science. 2012;**67**(1):101-108

[10] Sefton-Nash E, Catling DC. Hematitic concretions at Meridiani Planum, Mars: Their growth timescale and possible relationship with iron sulfates. Earth and Planetary Science Letters. 2008;**269**(3):366-376

[11] Squyres SW, Arvidson RE, Bell JF, Brückner J, Cabrol NA, Calvin W, et al. The Opportunity Rover's Athena science investigation at Meridiani Planum, Mars. Science. 2004;**306**(5702):1698

[12] Squyres SW, Arvidson RE, Bollen D, Bell JF, Brückner J, Cabrol NA, et al. Overview of the Opportunity Mars exploration Rover mission to Meridiani Planum: Eagle crater to purgatory ripple. Journal of Geophysical Research: Planets. 2006;(E12):111

[13] Squyres SW, Grotzinger JP, Arvidson RE, Bell JF, Calvin W, Christensen PR, et al. In Situ evidence for an ancient aqueous environment at Meridiani Planum, Mars. Science. 2004;**306**(5702):1709

[14] Squyres SW, Knoll AH, Arvidson RE, Ashley JW, Bell JF, Calvin WM, et al. Exploration of victoria crater by the mars rover Opportunity. Science. 2009;**324**(5930):1058

[15] Squyres SW, Knoll AH, Arvidson RE, Clark BC, Grotzinger JP, Jolliff BL, et al. Two years at Meridiani Planum: Results from the Opportunity Rover. Science. 2006;**313**(5792):1403

[16] Chan MA, Beitler B, Parry WT, Ormö J, Komatsu G. A possible

terrestrial analogue for haematite concretions on Mars. Nature. 2004;**429**:731

[17] Bowen BB, Benison KC, Oboh-Ikuenobe FE, Story S, Mormile MR. Active hematite concretion formation in modern acid saline lake sediments, Lake Brown, Western Australia. Earth and Planetary Science Letters. 2008;**268**(1):52-63

[18] Klingelhöfer G, Morris RV, Bernhardt B, Schröder C, Rodionov DS, de Souza PA, et al. Jarosite and hematite at Meridiani Planum from Opportunity's Mössbauer spectrometer. Science. 2004;**306**(5702):1740

[19] Knauth LP, Burt DM, Wohletz KH. Impact origin of sediments at the Opportunity landing site on Mars. Nature. 2005;**438**:1123

[20] Sexton MR, Elwood Madden ME, Swindle AL, Hamilton VE, Bickmore BR, Elwood Madden AS. Considering the formation of hematite spherules on Mars by freezing aqueous hematite nanoparticle suspensions. Icarus. 2017;**286**:202-211

[21] Arvidson RE, Anderson RC, Bartlett P, Bell JF, Christensen PR, Chu P, et al. Localization and physical property experiments conducted by Opportunity at Meridiani Planum. Science. 2004;**306**(5702):1730

[22] Misra AK, Acosta-Maeda TE, Scott ERD, Sharma SK. Possible mechanism for explaining the origin and size distribution of Martian hematite spherules. Planetary and Space Science. 2014;**92**:16-23

[23] Fairen AG, Dohm JM, Baker VR, Thompson SD, Mahaney WC, Herkenhoff KE, et al. Meteorites at Meridiani Planum provide evidence for significant amounts of surface and near-surface water on early Mars. Meteoritics & Planetary Science. 2011;**46**(12):1832-1841

[24] Schröder C, Rodionov DS, McCoy TJ, Jolliff BL, Gellert R, Nittler LR, et al. Meteorites on Mars observed with the Mars exploration Rovers. Journal of Geophysical Research: Planets. 2008;**113**(E06S22):1-19

[25] Clift R, Grace JR, Weber ME. Bubbles, Drops, and Particles. New York: Academic Press; 1978

[26] Kim HS, Kobayashi Y, Nagai K. Modeling of the surface tension of liquid Fe-P alloy by calculation of liquidus line in Fe-P binary system. Journal of Materials Research. 2006;**21**(6):1399-1408

[27] Beard KV, Chuang C. A new model for the equilibrium shape of raindrops. Journal of the Atmospheric Sciences. 1987;**44**(11):1509-1524

[28] Perlman H. Are Raindrops Shaped like Teardrops? USGS website. 2005. Available from: http://ga.water.usgs.gov/edu/raindropshape.html

[29] Ashley JW, Golombek MP, Christensen PR, Squyres SW, McCoy TJ, Schröder C, et al. Evidence for mechanical and chemical alteration of iron-nickel meteorites on Mars: Process insights for Meridiani Planum. Journal of Geophysical Research: Planets. 2011;**116**(E00F20):1-22

[30] Chappelow JE, Golombek MP. Event and conditions that produced the iron meteorite Block Island on Mars. Journal of Geophysical Research: Planets. 2010;**115**(E00F07):1-11

[31] Minitti ME, Lane MD, Bishop JL. A new hematite formation mechanism for Mars. Meteoritics & Planetary Science. 2005;**40**(1):55-69

[32] Cooper RF, Fanselow JB, Poker DB. The mechanism of oxidation of a basaltic glass: Chemical diffusion of network-modifying cations. Geochimica et Cosmochimica Acta. 1996;**60**(17):3253-3265

[33] Bohren CF, Olivero JJ. Evidence for haematite particles at 60 km altitude. Nature. 1984;**310**:216

[34] Brownlee DE. Cosmic dust: Collection and research. Annual Review of Earth and Planetary Sciences. 1985;**13**(1):147-173

[35] Brownlee DE, Bates B, Schramm L. The Leonard Award address presented 1996 July 25, Berlin, Germany: The elemental composition of stony cosmic spherules. Meteoritics & Planetary Science. 1997;**32**(2):157-175

[36] Taylor S, Brownlee DE. Cosmic spherules in the geologic record. Meteoritics. 1991;**26**(3):203-211

[37] Taylor S, Lever JH, Harvey RP. Numbers, types, and compositions of an unbiased collection of cosmic spherules. Meteoritics & Planetary Science. 2000;**35**(4):651-666

Chapter 2

Mineral Chemistry of Chalki Basalts in Northern Iraq and Their Petrological Significance

Mohsin M. Ghazal, Ali I. Al-Juboury and Sabhan M. Jalal

Abstract

Chalki basalts as a small body of volcanic rocks have green to grayish green color due to their nearly complete alteration to chlorite. The essential minerals of Chalki basalt to andesitic basalts are plagioclase (labradorite, An_{51-61}; andesine, An_{35} to An_{42}; and oligoclase, An_{22}). Moreover, there is sodic plagioclase (albite, $An_{0.1}$ to An_{04}) whose coexistence with the other more calcic plagioclase means that albitization had occurred. The other essential mineral is pyroxene (endiopside, en_{66-68} wo_{27-28} fs_{05-06}; and subcalcic augite, en_{72} wo_{14} fs_{14}). Olivine (Fo_{80-81}) is also present. According to the NiO content (0.11–0.12 wt%) in olivine grains, they are interpreted to be originated tectonically. The prevalent chlorite in all the samples is mainly diabantite and penninite, indicating chloritization after the ferromagnesian olivine and pyroxene. Serpentine (type lizardite and chrysotile) is also recorded as lesser alteration product after the forsteritic olivine. Rare secondary hornblende (type magnesiohornblende) is also found. The spinel group as accessory minerals is defined as magnetite, chromian magnetite, and chromian spinel giving the imprints of their metamorphic origin due to low temperature sub-sea metamorphism and also of alpine type.

Keywords: petrography, mineral chemistry, electron microprobe, Chalki basalts, Iraq

1. Introduction

The Paleozoic Chalki volcanic rocks crop out in a restricted part of the northern Thrust Zone of Iraq close to Iraqi-Turkish border and are considered as integral part of the upper part of Pirispiki Red Beds (late Devonian) Formation [1–3]. Chalki Formation was defined by Wetzel (unpublished report, 1952 in [1]) (which has no synonyms) who named it after the Chalki village (**Figure 1**).

Lithologically, they represent dull green and grayish green, red- and white-speckled, altered olivine-rich basaltic rocks (flow or intrusions) alternating with intercalations of bright red, ash-containing soft siltstones and shale.

They are undated and their origin is uncertain. Outcrop is typically not diagnostic. They are considered to be extrusive by [4] based on their identification of ash layers (not observed in this study). Petrographically, the bulk of the material consists of olivine basalts or fine-grained dolerites, with hematite-magnetite-rimmed

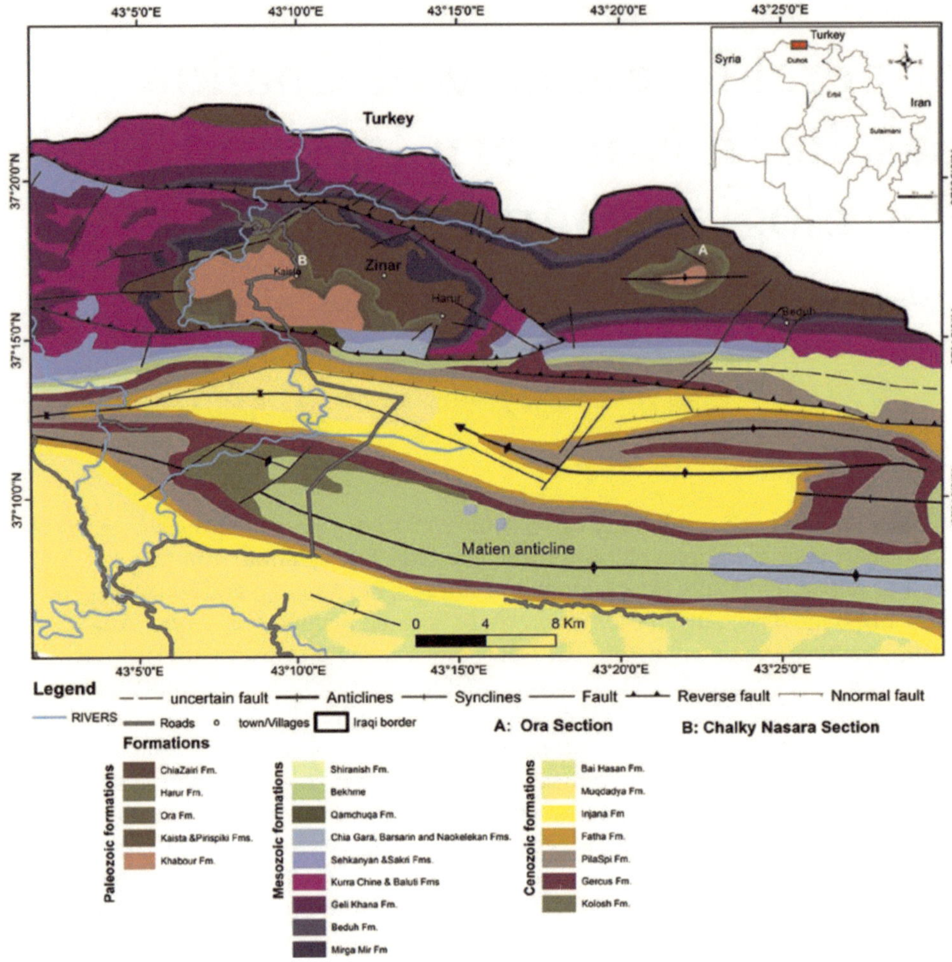

Figure 1.
Geological map of northern Iraq showing the Paleozoic succession including Chalki volcanics within Pirispiki formation in Ora section (A) and the studied section (Chalki Nasara, section B) (modified after [8]).

pseudomorphs, in chlorite, replacing the olivine. There are albitized plagioclase laths and considerable amounts of chlorite and ankeritic carbonates in the groundmass. Locally the basalts are crossed by numerous veins of white ankerite with fibrous chalcedony.

The Chalki volcanic rocks probably had suffered severe sub-sea alterations. Therefore, this work aims to study mineral chemistry using electron probe microanalyzer (EPMA) in order to distinguish the various phases of minerals resulted from the alteration on basaltic rocks of the Chalki volcanics and interpret their petrologic significance.

2. Geologic setting

The Chalki volcanics located near Kaista (Khabour valley, Amadia district, N. Iraq) occur as basalt intercalations, of 2–5 m in thickness within the Pirispiki Formation. The type section lies at Chalki village (**Figure 1**) in which the basaltic beds are associated with ash-containing shales and siltstones occupying most of the uppermost 20 m of the section. The type section is 16 m thick in aggregate [4].

Sharland et al. [5] interpreted the Chalki volcanics to represent back-arc rift volcanics associated with the initiation of subduction along the Tethyan margin of the Arabian Plate. They interpreted the initiation of subduction to have caused the so-called Hercynian orogeny in the late Devonian times. The age of the "Hercynian orogeny" in the Arabian Plate has been reported to range from pre-Late Devonian to middle Carboniferous [6].

Subduction along the southern margin of the Palaeo-Tethys is supported by the occurrence of Devonian-Carboniferous volcanic and metamorphic rocks found in the Kuh-Sefid area of the Sanandaj-Sirjan Zone [7]. However, the central part of the Arabian plate was probably not significantly affected by this subduction.

3. Materials and methodology

The petrography of 15 samples collected from Chalki volcanics section at Chalki Nasara section (**Figure 1**) was determined using petrographic microscope at the Earth Science Department of Mosul University, Iraq. A Swift Polarized microscope is used in the petrographic description.

Electron microprobe analyses of minerals were performed using a Cameca SX-50 in the Department of Geology and Geophysics at the University of Utah, USA. Analyses were conducted using PC1, TAP, PET, and LiF crystals on four wavelength-dispersive spectrometers, with an accelerating voltage of 15 keV, a beam current of 30 nA, and a spot size of 10 mm. Peak intensities were measured for 20 s and background intensities for 10 s on both sides of the analytical peaks. Na was measured first, and analytical times were reduced by half in order to minimize sodium loss under the electron beam. The analytical routine for feldspars included Si, Al, Fe, Ca, Sr, Ba, Na, and K, and a separate analytical routine for mafic and other minerals added Ti, Cr, Mn, Ni, Zn, Mg, F, and Cl (K, Sr and Ba excluded). Mineral standards include fluorite (F-Kα), tugtupite (Cl-Kα), sanidine (Si-Kα, Al-Kα, K-Kα), albite (Na-Kα), plagioclase (Ca-Kα), barite (Ba-Lα), celestite (Sr-Lα), chrome diopside (Mg-Kα), hematite (Fe-Kα), rutile (Ti-Kα), rhodonite (Mn-Kα), chromite (Cr-Kα), nickel silicide (Ni-Kα), and zinc sulfide (Zn-Kβ). Rounds of standard analyses were performed prior to and after the suite of thin sections. Concentrations are calculated using the PAP matrix correction factors. Correction for "excess" F by interference of the Fe-Lα peak with F-Kα peak was accomplished by measuring an F-free Fe-bearing standard (hematite) and calculating a correction factor of 0.029.

4. Results

4.1 Petrography

The volcanic rocks are olivine-basalt, sometimes doleritic, or even andesitic-basalt. They are of greenish to greenish gray color due to high chloritization after olivine. They are associated with little phyllites and pyroclasts of volcano-sedimentary rocks. They contain also veins of carbonate minerals and quartz.

Texturally, the rocks of Chalki volcanics are porphyritic basaltic in general, cut by microscopic veins of carbonate minerals (calcite) in addition to minute quartz veins.

Mineralogically, the pseudomorphs after olivine phenocrysts are identified microscopically, such as chlorite, iddingsite, and iron oxides (**Figure 2**). Another essential mineral is plagioclase (sometimes albitized), which sometimes slightly altered to sericite and kaolinite. No individual pyroxene grains have been identified, but they probably present as fine grains in the groundmass.

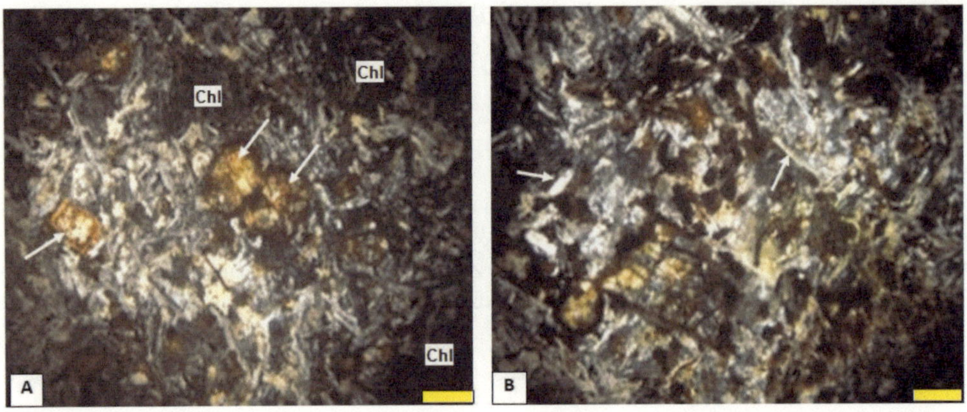

Figure 2.
Representative photomicrograph for Chalki basalt showing (A) dumpy prismatic olivine with rim alteration to black iron oxide and reddish brown iddingsite (arrows), chlorite (Chl), and white fine laths of plagioclase. (B) Plagioclase laths (arrows), altered olivine to iddingsite, iron oxide. Under-crossed nicols. Scale bar is 0.2 mm.

4.2 Mineral chemistry

4.2.1 Plagioclase

The calcic plagioclase (labradorite, An_{51} to An_{61}) is the essential mineral of Chalki basalt resembling the basaltic type. Some grains are less calcic (andesine, An_{35} to An_{42}). Moreover, there is sodic plagioclase (albite, $An_{0.1}$ to An_{04}) as a result of albitization process (**Table 1** and **Figure 3**).

CV-32								
Wt%	1	2	3	4	5	6	7	8
SiO_2	54.59	56.20	57.15	66.91	65.28	67.76	58.82	55.16
Al_2O_3	27.44	26.06	26.25	19.41	20.81	19.65	25.04	27.56
FeO^*	0.59	0.92	0.87	0.91	0.56	0.92	0.77	0.77
CaO	10.32	8.56	8.49	0.34	0.88	0.15	7.25	10.15
SrO	0.00	0.00	0.00	0.00	0.00	0.00	0.00	0.03
BaO	0.00	0.01	0.00	0.00	0.00	0.00	0.00	0.00
Na_2O	5.33	6.24	6.07	10.70	10.20	11.23	6.63	5.16
K_2O	0.35	0.24	0.50	0.10	0.91	0.13	0.58	0.27
Total	98.6	98.2	99.3	98.4	98.6	99.8	99.1	99.1
apfu								
Si	2.501	2.575	2.587	2.979	2.914	2.976	2.657	2.511
Al	1.481	1.407	1.400	1.018	1.095	1.017	1.333	1.478
Fe^{2+}	0.023	0.035	0.033	0.034	0.021	0.034	0.029	0.029
Ca	0.507	0.420	0.412	0.016	0.042	0.007	0.351	0.495
Sr	0.000	0.000	0.000	0.000	0.000	0.000	0.000	0.000
Ba	0.000	0.000	0.000	0.000	0.000	0.000	0.000	0.000
Na	0.473	0.554	0.533	0.924	0.883	0.956	0.580	0.456
K	0.021	0.014	0.029	0.006	0.052	0.007	0.034	0.016

CV-32								
Wt%	**1**	**2**	**3**	**4**	**5**	**6**	**7**	**8**
Ab	47.30	56.07	54.75	97.67	90.34	98.52	60.16	47.17
An	50.63	42.48	42.31	1.73	4.33	0.71	36.35	51.21
Or	2.07	1.45	2.94	0.60	5.33	0.77	3.49	1.61
	Labradorite	Andesine	Andesine	Albite	Albite	Albite	Andesine	Labradorite

CV-25						
Wt%	**1**	**2**	**3**	**4**	**5**	**6**
SiO_2	57.80	53.20	52.54	62.32	52.91	57.66
Al_2O_3	25.46	28.77	29.17	22.97	29.20	26.26
FeO*	1.20	0.68	0.86	0.71	0.67	0.55
CaO	7.52	11.68	12.08	4.71	11.64	8.36
SrO	0.00	0.02	0.00	0.06	0.00	0.00
BaO	0.00	0.00	0.00	0.05	0.01	0.03
Na_2O	6.60	4.61	4.15	8.64	4.29	6.32
K_2O	0.56	0.19	0.33	0.63	0.52	0.52
Total	99.1	99.1	99.1	100.1	99.2	99.7
apfu						
Si	2.621	2.433	2.408	2.774	2.419	2.597
Al	1.361	1.551	1.576	1.205	1.573	1.394
Fe^{2+}	0.046	0.026	0.033	0.026	0.026	0.021
Ca	0.365	0.572	0.593	0.225	0.570	0.404
Sr	0.000	0.000	0.000	0.000	0.000	0.000
Ba	0.000	0.000	0.000	0.000	0.000	0.000
Na	0.580	0.408	0.369	0.745	0.380	0.552
K	0.032	0.011	0.019	0.036	0.030	0.030
Ab	59.35	41.18	37.58	74.12	38.76	56.01
An	37.35	57.72	60.45	22.34	58.13	40.94
Or	3.30	1.10	1.97	3.54	3.11	3.05
	Andesine	Labradorite	Labradorite	Oligoclase	Labradorite	Andesine

CV-37					
Wt%	**1**	**2**	**3**	**4**	**5**
SiO_2	57.32	54.96	58.70	56.93	55.49
Al_2O_3	25.85	27.08	24.96	24.40	27.90
FeO*	1.00	1.19	0.87	1.65	0.98
CaO	8.10	9.94	7.05	6.76	10.06
SrO	0.00	0.00	0.00	0.00	0.00
BaO	0.00	0.00	0.03	0.02	0.01
Na_2O	6.13	5.28	6.79	6.25	5.46
K_2O	0.48	0.27	0.61	0.61	0.39
Total	98.9	98.7	99.0	96.6	100.3

CV-37					
Wt%	**1**	**2**	**3**	**4**	**5**
apfu					
Si	2.604	2.516	2.656	2.647	2.502
Al	1.384	1.461	1.331	1.337	1.482
Fe^{2+}	0.038	0.046	0.033	0.064	0.037
Ca	0.394	0.487	0.342	0.337	0.486
Sr	0.000	0.000	0.000	0.000	0.000
Ba	0.000	0.000	0.000	0.000	0.000
Na	0.540	0.469	0.595	0.563	0.477
K	0.028	0.016	0.035	0.036	0.022
Ab	56.14	48.23	61.26	60.16	48.43
An	40.99	50.14	35.14	35.98	49.31
Or	2.87	1.64	3.61	3.86	2.26
	Andesine	Labradorite	Andesine	Andesine	Labradorite

CV-41									
Wt%	**1**	**2**	**3**	**4**	**5**	**6**	**7**	**8**	**9**
SiO_2	68.13	67.84	68.02	67.77	68.20	66.15	67.24	67.39	67.87
Al_2O_3	19.45	19.34	19.25	19.59	19.56	20.09	19.51	19.46	19.59
FeO*	0.47	0.56	0.77	0.95	0.54	1.55	0.77	0.71	0.63
CaO	0.04	0.04	0.03	0.10	0.07	0.41	0.32	0.08	0.12
SrO	0.00	0.00	0.00	0.00	0.00	0.00	0.00	0.00	0.00
BaO	0.00	0.01	0.00	0.00	0.00	0.37	0.00	0.00	0.00
Na_2O	11.16	11.13	11.24	11.18	11.43	10.42	11.04	11.09	11.14
K_2O	0.03	0.04	0.02	0.07	0.05	0.29	0.07	0.17	0.09
Total	99.3	99.0	99.3	99.7	99.9	99.3	98.9	98.9	99.4
apfu									
Si	2.996	2.995	2.996	2.980	2.988	2.940	2.977	2.983	2.985
Al	1.008	1.006	0.999	1.015	1.010	1.053	1.018	1.015	1.016
Fe^{2+}	0.017	0.021	0.028	0.035	0.020	0.058	0.028	0.026	0.023
Ca	0.002	0.002	0.001	0.005	0.003	0.020	0.015	0.004	0.006
Sr	0.000	0.000	0.000	0.000	0.000	0.000	0.000	0.000	0.000
Ba	0.000	0.000	0.000	0.000	0.000	0.001	0.000	0.000	0.000
Na	0.951	0.953	0.959	0.953	0.971	0.898	0.948	0.952	0.950
K	0.002	0.002	0.001	0.004	0.003	0.016	0.004	0.010	0.005
Ab	99.59	99.59	99.72	99.11	99.39	96.15	98.03	98.61	98.90
An	0.21	0.19	0.14	0.50	0.32	2.10	1.59	0.37	0.60
Or	0.21	0.22	0.13	0.39	0.30	1.75	0.39	1.02	0.51
	Albite	Albite	Albite	Albite	Albite	Albite	Albite	Albite	Albite

Table 1.
Chemical composition (wt%) and atom per formula unit (apfu) of plagioclase on the basis of 8 O (samples CV-32, CV-25, CV-37, CV-41).

4.2.2 Olivine

Chalki olivine is forsteritic (Fo_{80} to Fo_{81}) and consequently high Mg# (80–81) (**Table 2**). It contains NiO (0.11–0.12 wt%), which is plotted against (Fo%) revealing its tectonic origin (**Figure 4**).

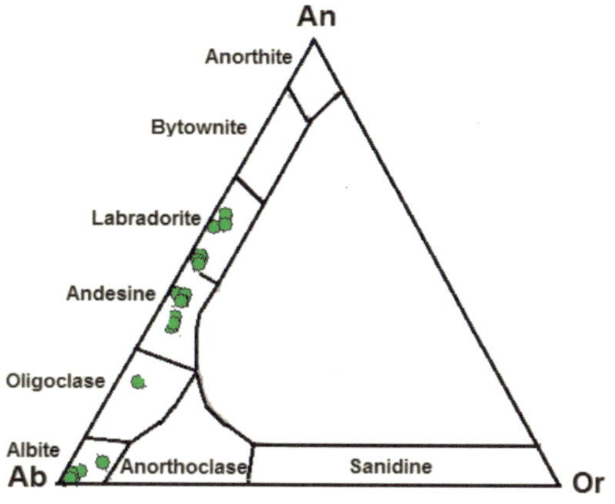

Figure 3.
An-Ab-Or plot for Chalki basalt plagioclase showing the composition variation from labradorite, andesine, oligoclase, to albite (modified from [9]).

Wt%	1	2	3	4
SiO_2	39.29	40.01	38.52	39.34
TiO_2	0.02	0.04	0.02	0.04
Al_2O_3	0.19	0.24	0.19	0.24
Cr_2O_3	0.02	0.06	0.02	0.06
FeO*	17.20	18.01	16.86	17.71
MnO	0.37	0.48	0.36	0.47
NiO	0.12	0.12	0.12	0.11
ZnO	0.54	0.00	0.53	0.00
MgO	43.03	41.35	42.19	40.66
CaO	0.01	0.64	0.01	0.63
Na_2O	0.04	0.01	0.04	0.01
F	0.05	0.07	0.05	0.07
Cl	0.00	0.00	0.00	0.00
Total	100.88	101.03	98.90	99.34
apfu				
Si	0.992	1.010	0.992	1.010
Al	0.006	0.007	0.006	0.007
Ti	0.000	0.001	0.000	0.001
Cr	0.000	0.001	0.000	0.001
Fe	0.363	0.380	0.363	0.380
Mn	0.008	0.010	0.008	0.010

Wt%	1	2	3	4
Mg	1.620	1.556	1.620	1.556
Ca	0.000	0.017	0.000	0.017
Na	0.002	0.001	0.002	0.001
K	0.000	0.000	0.000	0.000
Ni	0.001	0.001	0.001	0.001
Zn	0.004	0.000	0.004	0.000
Fo	81.361	79.946	81.361	79.946
Fa	18.639	20.054	18.639	20.054
Mg#	81.686	80.367	81.686	80.367

Table 2.
Chemical composition (wt%) and atom per formula unit (apfu) of olivine on the basis of 4 O (sample CV-67).

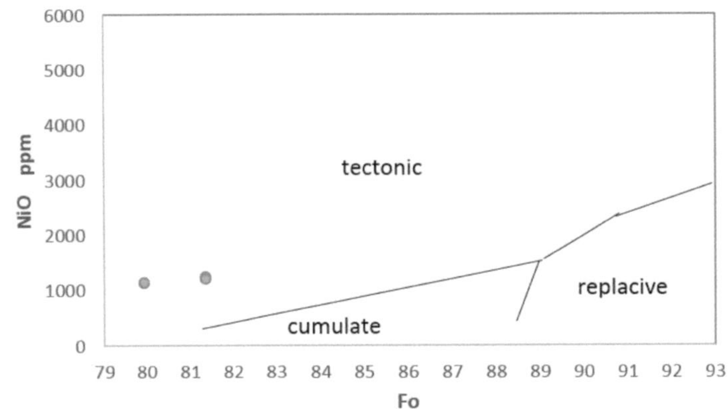

Figure 4.
Fo vs. NiO for Chalki basalt olivines shown as type tectonic (adapted from [10]).

4.2.3 Pyroxene

The pyroxene of Chalki basalt is endiopside to subcalcic augite as shown in **Table 3** and **Figure 5**.

The chemical formula of pyroxene is XYZ_2O_6 [11].

4.2.4 Hornblende

The rare mineral in Chalki basalt is hornblende, whose composition resembles the magnnesiohornblende as shown in **Figure 6** after representation of its chemical composition given in **Table 4**.

The general formula of amphiboles is $W_{(0\text{–}1)}X_2Y_5Z_8O_{22}(OH,F)_2$ [11].

4.2.5 Chlorite

The chlorite of Chalki basalt is the predominant secondary mineral in all samples giving the greenish color of the hand specimen samples. The huge number of

Wt%	1	2	3	4	5
SiO_2	56.54	56.43	56.53	57.31	50.47
TiO_2	0.08	0.07	0.03	0.06	0.05
Al_2O_3	1.92	2.05	1.58	1.11	0.88
Cr_2O_3	0.09	0.12	0.19	0.08	0.11
FeO^*	3.28	3.22	3.12	3.39	9.99
MnO	0.13	0.03	0.08	0.09	0.21
NiO	0.05	0.05	0.03	0.02	0.04
ZnO	0.21	0.00	0.00	0.00	0.02
MgO	22.27	22.33	22.91	22.75	29.19
CaO	13.17	13.03	12.79	12.75	8.18
Na_2O	0.23	0.29	0.19	0.10	0.12
F	0.01	0.00	0.00	0.00	0.05
Cl	0.00	0.01	0.01	0.01	0.01
Total	98.0	97.6	97.5	97.7	99.3
apfu					
Si	2.028	2.027	2.033	2.055	1.850
Al^{iv}	−0.028	−0.027	−0.033	−0.055	0.150
Al^{vi}	0.109	0.114	0.100	0.101	−0.112
Ti	0.002	0.002	0.001	0.001	0.001
Cr	0.002	0.003	0.006	0.002	0.003
Fe^{2+}	−0.314	−0.316	−0.335	−0.324	−0.495
Mn	0.004	0.001	0.000	0.003	0.006
Mg	1.191	1.196	1.229	1.216	1.595
Ni	0.002	0.002	0.001	0.000	0.001
Zn	0.006	0.000	0.000	0.000	0.001
Fe^{2+}	0.412	0.413	0.429	0.426	0.801
Ca	0.506	0.502	0.493	0.490	0.321
Na	0.016	0.020	0.014	0.007	0.008
K	0.000	0.000	0.000	0.000	0.000
Z	2.000	2.000	2.000	2.000	2.000
Y	1.002	1.002	1.001	1.000	1.001
X	0.934	0.935	0.936	0.923	1.131
Xmg	0.924	0.925	0.929	0.923	0.839
fs	5.478	5.386	5.177	5.626	13.776
en	66.322	66.659	67.675	67.277	71.777
wo	28.199	27.955	27.148	27.097	14.447

Table 3.
Chemical composition (wt%) and atom per formula unit (apfu) of pyroxene on the basis of 6 O (sample CV-67).

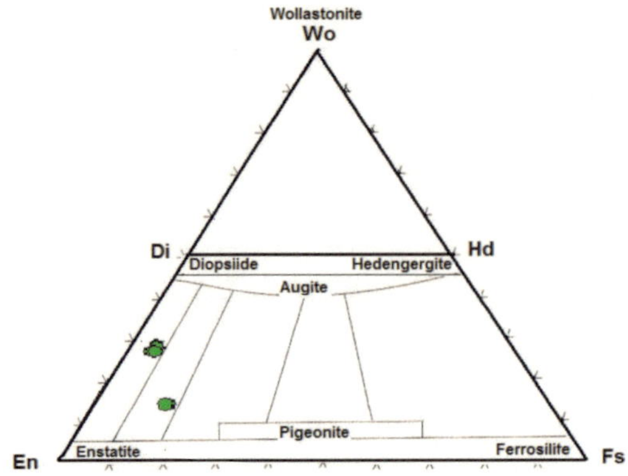

Figure 5.
En-Fs-Wo triangular diagram for Chalki basalt pyroxene shown as endiopside and subcalcic augite (after [11]).

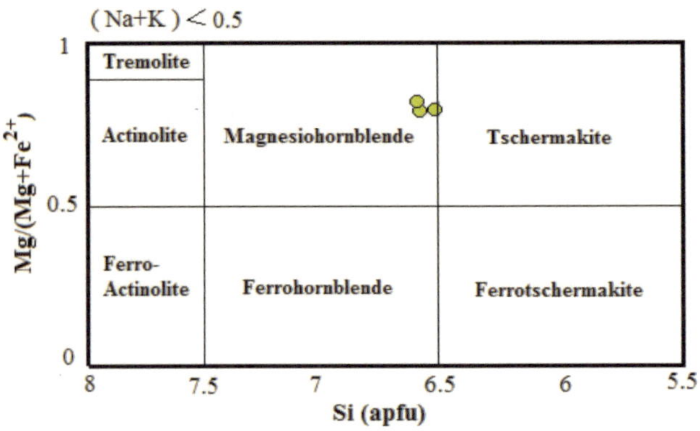

Figure 6.
Si vs. Mg/(Mg + Fe^{2+}) diagram for Chalki basalt amphiboles shown as magnesiohornblende (after [12]).

Wt%	1	2	3
SiO_2	45.77	44.9	45.8
Al_2O_3	11.09	11.45	11.45
TiO_2	0.91	1.1	0.77
Cr_2O_3	1.08	0.96	0.67
FeO*	7.03	7.99	8.02
MnO	0.06	0.03	0.01
MgO	17.04	16.55	16.87
CaO	12.14	12.4	12.66
Na_2O	1.45	0.88	0.6
Total	96.6	96.3	96.9
(apfu)			
Si	6.600	6.522	6.592
Al^{iv}	1.400	1.478	1.408
Al^{vi}	0.483	0.482	0.534

Wt%	**1**	**2**	**3**
Ti	0.099	0.120	0.083
Cr	0.123	0.110	0.076
Fe^{2+}	0.627	0.701	0.685
Mn	0.007	0.004	0.001
Mg	3.662	3.584	3.620
Fe^{2+}	0.221	0.270	0.280
Ca	1.874	1.930	1.952
Na	−0.095	−0.199	−0.232
K	0	0	0
Z	8	8	8
Y	5	5	5
X	2	2	2
W	0	0	0
Mg#	0.812	0.787	0.789

Table 4.
Chemical composition (wt%) and atom per formula unit (apfu) of hornblende on the basis of 23 O (sample CV-67).

CV-32							**CV-67**	
Wt%	**1**	**2**	**3**	**4**	**5**	**6**	**1**	**2**
SiO_2	32.02	32.76	33.26	34.05	34.88	35.47	32.24	30.38
TiO_2	0.09	0.00	0.04	0.05	0.00	0.02	0.02	0.03
Al_2O_3	13.58	14.80	13.44	13.81	13.75	13.32	13.90	18.68
Cr_2O_3	0.03	0.01	0.04	0.01	0.03	0.00	1.14	0.16
FeO*	17.25	14.75	16.25	14.52	13.18	13.46	7.17	6.04
MnO	0.11	0.06	0.05	0.02	0.03	0.03	0.03	0.06
NiO	0.08	0.09	0.09	0.07	0.09	0.07	0.05	0.10
ZnO	0.00	0.23	0.07	0.37	0.24	0.11	0.06	0.00
MgO	21.66	21.64	22.48	22.76	23.76	23.54	31.03	30.76
CaO	0.48	0.47	0.38	0.39	0.31	0.32	0.26	0.03
Na_2O	0.02	0.05	0.06	0.04	0.07	0.12	0.01	0.05
F	0.07	0.04	0.06	0.04	0.08	0.10	0.08	0.03
Cl	0.01	0.00	0.01	0.00	0.01	0.03	0.02	0.01
Total	85.4	84.9	86.2	86.1	86.4	86.6	86.0	86.3
Apfu								
Si	6.611	6.689	6.742	6.836	6.916	7.020	6.309	5.861
Al^{iv}	1.389	1.311	1.258	1.164	1.084	0.980	1.691	2.139
Al^{vi}	1.914	2.251	1.954	2.103	2.128	2.126	1.513	2.108
Ti	0.013	0.000	0.007	0.008	0.000	0.003	0.003	0.005
Cr	0.004	0.002	0.006	0.002	0.004	0.000	0.177	0.024

CV-32							**CV-67**	
Wt%	**1**	**2**	**3**	**4**	**5**	**6**	**1**	**2**
Fe	2.979	2.518	2.755	2.437	2.184	2.227	1.174	0.974
Mn	0.020	0.010	0.008	0.004	0.006	0.004	0.006	0.010
Ni	0.013	0.015	0.015	0.012	0.015	0.012	0.007	0.016
Zn	0.000	0.035	0.010	0.054	0.035	0.017	0.008	0.000
Mg	6.667	6.587	6.794	6.811	7.024	6.945	9.054	8.848
Ca	0.107	0.103	0.084	0.083	0.066	0.068	0.054	0.005
Na	0.008	0.018	0.022	0.015	0.026	0.046	0.003	0.019
CV-25								
Wt%	**1**	**2**	**3**	**4**	**5**	**6**	**7**	
SiO_2	23.44	25.96	27.76	31.42	33.24	36.69	37.34	
TiO_2	0.72	0.24	0.33	0.19	0.05	0.00	0.04	
Al_2O_3	11.42	13.15	13.84	15.31	16.19	19.58	21.16	
Cr_2O_3	0.01	0.11	0.04	0.01	0.00	0.03	0.08	
FeO^*	34.69	30.38	26.27	17.29	14.32	12.79	11.40	
MnO	0.03	0.04	0.04	0.07	0.09	0.01	0.00	
NiO	0.08	0.08	0.10	0.11	0.09	0.08	0.06	
ZnO	0.12	0.00	0.10	0.11	0.00	0.34	0.33	
MgO	17.16	18.73	19.86	22.71	23.52	20.32	18.84	
CaO	0.11	0.13	0.12	0.11	0.24	0.27	0.26	
Na_2O	0.04	0.02	0.04	0.01	0.04	0.07	0.06	
F	0.02	0.04	0.07	0.11	0.11	0.10	0.10	
Cl	0.02	0.02	0.01	0.02	0.01	0.01	0.01	
Total	87.9	88.9	88.6	87.5	87.9	90.3	89.7	
apfu								
Si	5.332	5.624	5.868	6.334	6.524	6.862	6.947	
Al^{iv}	2.668	2.376	2.132	1.666	1.476	1.138	1.053	
Al^{vi}	0.394	0.981	1.315	1.972	2.269	3.179	3.587	
Ti	0.124	0.039	0.053	0.029	0.007	0.000	0.005	
Cr	0.002	0.018	0.007	0.002	0.001	0.005	0.012	
Fe	6.598	5.504	4.642	2.914	2.350	2.000	1.773	
Mn	0.005	0.007	0.007	0.013	0.014	0.002	0.000	
Ni	0.014	0.014	0.017	0.018	0.013	0.011	0.009	
Zn	0.020	0.000	0.016	0.017	0.000	0.048	0.046	
Mg	5.818	6.050	6.260	6.826	6.883	5.666	5.226	
Ca	0.027	0.030	0.027	0.025	0.049	0.053	0.051	
Na	0.018	0.010	0.017	0.005	0.017	0.026	0.023	
CV-37								
Wt%	**1**	**2**	**3**	**4**	**5**	**6**	**7**	**8**
SiO_2	32.18	32.68	32.69	32.85	32.87	32.95	32.96	33.09

CV-37								
Wt%	**1**	**2**	**3**	**4**	**5**	**6**	**7**	**8**
TiO_2	0.02	0.04	0.00	0.05	0.09	0.01	0.06	0.00
Al_2O_3	15.68	15.55	15.56	15.66	13.92	15.62	15.48	15.16
Cr_2O_3	0.13	0.18	0.09	0.08	0.00	0.15	0.02	0.04
FeO*	14.75	14.29	14.47	14.77	17.19	14.96	15.08	14.87
MnO	0.01	0.03	0.04	0.03	0.09	0.04	0.04	0.04
NiO	0.05	0.11	0.06	0.11	0.07	0.06	0.06	0.12
ZnO	0.00	0.18	0.00	0.00	0.43	0.39	0.26	0.18
MgO	23.81	24.67	24.61	24.21	22.20	24.20	23.55	23.70
CaO	0.12	0.20	0.12	0.15	0.39	0.16	0.18	0.20
Na_2O	0.04	0.02	0.03	0.01	0.01	0.03	0.05	0.03
F	0.06	0.10	0.04	0.11	0.05	0.09	0.12	0.10
Cl	0.01	0.01	0.01	0.01	0.00	0.01	0.01	0.01
Total	86.9	88.1	87.7	88.0	87.3	88.7	87.9	87.5
apfu								
Si	6.421	6.429	6.429	6.465	6.629	6.459	6.516	6.558
Al^{iv}	1.579	1.571	1.571	1.535	1.371	1.541	1.484	1.442
Al^{vi}	2.110	2.033	2.033	2.097	1.938	2.067	2.123	2.098
Ti	0.003	0.006	0.006	0.008	0.013	0.001	0.008	0.000
Cr	0.020	0.029	0.029	0.013	0.000	0.023	0.003	0.007
Fe	2.462	2.351	2.351	2.430	2.898	2.451	2.493	2.464
Mn	0.001	0.005	0.005	0.005	0.015	0.006	0.007	0.006
Ni	0.008	0.018	0.018	0.017	0.011	0.010	0.010	0.019
Zn	0.000	0.027	0.027	0.000	0.065	0.056	0.039	0.027
Mg	7.084	7.235	7.235	7.102	6.675	7.070	6.941	7.000
Ca	0.026	0.042	0.042	0.032	0.085	0.033	0.037	0.043
Na	0.016	0.006	0.006	0.004	0.005	0.012	0.018	0.010
CV-41								
Wt%	**1**	**2**	**3**	**4**	**5**	**6**	**7**	**8**
SiO_2	33.06	32.54	32.39	32.62	32.58	32.06	31.82	33.02
TiO_2	0.05	0.04	0.04	0.04	0.01	0.01	0.00	0.00
Al_2O_3	14.98	15.00	15.02	15.04	15.26	15.40	15.44	15.45
Cr_2O_3	0.09	0.21	0.20	0.08	0.08	0.03	0.02	0.05
FeO*	16.38	16.66	16.62	16.96	15.84	16.38	17.11	16.05
MnO	0.02	0.06	0.01	0.02	0.05	0.09	0.11	0.06
NiO	0.02	0.03	0.05	0.11	0.09	0.11	0.06	0.10
ZnO	0.00	0.00	0.06	0.17	0.37	0.02	0.05	0.61
MgO	22.77	22.67	22.57	22.45	22.46	22.28	21.63	22.80
CaO	0.20	0.19	0.20	0.21	0.22	0.29	0.61	0.19
Na_2O	0.03	0.02	0.03	0.03	0.05	0.02	0.00	0.01

CV-41								
Wt%	**1**	**2**	**3**	**4**	**5**	**6**	**7**	**8**
F	0.06	0.07	0.09	0.07	0.00	0.09	0.09	0.06
Cl	0.01	0.02	0.02	0.01	0.00	0.01	0.02	0.00
Total	87.7	87.5	87.3	87.8	87.0	86.8	87.0	88.4
apfu								
Si	6.579	6.512	6.501	6.519	6.532	6.465	6.440	6.529
Al^{iv}	1.421	1.488	1.499	1.481	1.468	1.535	1.560	1.471
Al^{vi}	2.092	2.049	2.052	2.061	2.139	2.126	2.122	2.130
Ti	0.007	0.006	0.006	0.006	0.002	0.001	0.000	0.000
Cr	0.014	0.033	0.032	0.012	0.013	0.005	0.004	0.008
Fe	2.726	2.787	2.789	2.834	2.655	2.763	2.894	2.653
Mn	0.004	0.011	0.002	0.003	0.008	0.016	0.020	0.010
Ni	0.002	0.004	0.009	0.018	0.014	0.018	0.010	0.016
Zn	0.000	0.000	0.008	0.025	0.054	0.003	0.007	0.088
Mg	6.755	6.763	6.754	6.688	6.713	6.701	6.527	6.719
Ca	0.042	0.040	0.043	0.044	0.047	0.063	0.132	0.041
Na	0.010	0.007	0.012	0.013	0.020	0.009	0.001	0.005

CV-41								
Wt%	**9**	**10**	**11**	**12**	**13**	**14**	**15**	**16**
SiO_2	32.62	31.97	32.53	31.62	31.35	34.42	32.19	30.87
TiO_2	1.22	0.08	0.00	0.04	0.01	0.05	0.01	0.01
Al_2O_3	15.50	15.56	16.13	16.15	16.31	16.46	16.71	17.19
Cr_2O_3	0.14	0.13	0.11	0.12	0.15	0.10	0.17	0.18
FeO*	15.25	16.67	14.47	14.18	14.59	13.95	14.87	14.53
MnO	0.04	0.05	0.04	0.04	0.02	0.03	0.03	0.07
NiO	0.05	0.11	0.19	0.09	0.11	0.11	0.11	0.06
ZnO	0.25	0.01	0.10	0.00	0.14	0.02	0.28	0.32
MgO	23.17	21.61	24.06	21.94	22.74	20.25	23.35	22.42
CaO	0.17	0.40	0.11	0.21	0.13	0.60	0.13	0.12
Na_2O	0.01	0.02	0.05	0.01	0.03	0.93	0.05	0.04
F	0.07	0.11	0.10	0.00	0.08	0.08	0.08	0.01
Cl	0.02	0.01	0.00	0.02	0.00	0.01	0.02	0.01
Total	88.5	86.7	87.9	84.4	85.7	87.0	88.0	85.8
apfu								
Si	6.414	6.467	6.410	6.469	6.352	6.803	6.353	6.243
Al^{iv}	1.586	1.533	1.590	1.531	1.648	1.197	1.647	1.757
Al^{vi}	2.006	2.175	2.156	2.362	2.246	2.637	2.239	2.342
Ti	0.181	0.012	0.000	0.006	0.001	0.008	0.002	0.001
Cr	0.021	0.020	0.017	0.019	0.023	0.016	0.026	0.029

CV-41								
Wt%	9	10	11	12	13	14	15	16
Fe	2.508	2.819	2.384	2.425	2.471	2.306	2.454	2.458
Mn	0.007	0.008	0.007	0.006	0.004	0.004	0.004	0.012
Ni	0.008	0.017	0.030	0.014	0.018	0.018	0.018	0.010
Zn	0.036	0.002	0.015	0.000	0.021	0.003	0.040	0.048
Mg	6.792	6.514	7.067	6.689	6.870	5.967	6.869	6.761
Ca	0.036	0.086	0.024	0.045	0.029	0.126	0.027	0.025
Na	0.005	0.008	0.018	0.004	0.013	0.357	0.019	0.015

Table 5.
Chemical composition (wt%) and atom per formula unit (apfu) of chlorite on the basis of 28 O (samples CV-32, CV-67, CV-25, CV-37, CV-41).

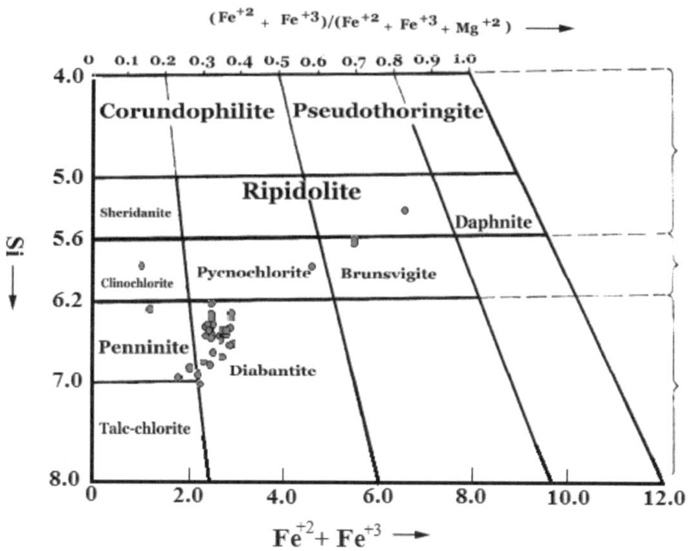

Figure 7.
Chalki basalt chlorite shown as mainly diabantite and less penninite (after [13]).

analysis for many samples as given in **Table 5** reveals the prevalence of diabantite with less abundant penninite, in addition to few grains of clinochlorite, pyenochlorite, and ripidolite (**Figure 7**).

4.2.6 Spinel group

The chemical data (**Table 6**) and their representation (**Figure 8**) for the few measured points of spinel group display that they are magnetite, chromian magnetite, and chromian spinel. Moreover, the plot of their Cr# against Mg# (**Figure 9**) gives an important and clear sign for their metamorphic origin as a result of low-grade metamorphism and of alpine type.

S 25				S 41	S 67	
Wt%	**1**	**2**	**3**	**1**	**1**	**2**
SiO_2	0.25	0.08	1.08	0.12	0.11	0.19
TiO_2	10.93	0.61	1.72	0.50	0.81	0.97
Al_2O_3	4.19	36.74	20.12	41.11	0.42	0.26
Cr_2O_3	21.07	22.96	30.26	23.03	2.03	2.50
FeO*	55.68	28.02	36.84	17.66	89.91	89.06
MnO	0.01	0.00	0.00	0.00	0.09	0.08
NiO	0.10	0.13	0.08	0.20	0.19	0.17
ZnO	1.81	0.00	0.29	0.03	0.44	0.35
MgO	0.17	10.87	5.49	15.45	0.43	0.38
CaO	0.02	0.00	0.02	0.52	0.02	0.01
Na_2O	0.03	0.03	0.00	0.02	0.01	0.00
F	0.00	0.02	0.00	0.10	0.00	0.00
Cl	0.00	0.00	0.01	0.01	0.00	0.00
Total	94.3	99.5	95.9	98.8	94.5	94.0
apfu						
Si	0.0101	0.0024	0.0374	0.0035	0.0053	0.0095
Al	0.2022	1.3019	0.8213	1.3889	0.0242	0.0154
Ti	0.3367	0.0138	0.0448	0.0108	0.0301	0.0362
Cr	0.6827	0.5457	0.8287	0.5219	0.0793	0.0978
$\sum$	*1.2317*	*1.8639*	*1.7321*	*1.9250*	*0.1388*	*0.1590*
Fe	1.9083	0.7045	1.0670	0.4233	3.7136	3.6864
Mn	0.0004	0.0000	0.0000	0.0000	0.0037	0.0035
Ni	0.0034	0.0031	0.0022	0.0047	0.0075	0.0069
Zn	0.0547	0.0000	0.0074	0.0007	0.0160	0.0127
Mg	0.0106	0.4875	0.2834	0.6601	0.0315	0.0284
Ca	0.0007	0.0001	0.0008	0.0160	0.0013	0.0007
Na	0.0020	0.0017	0.0000	0.0010	0.0012	0.0000
$\sum$	*1.9801*	*1.1969*	*1.3607*	*1.1058*	*3.7747*	*3.7387*
Fe^{2+*}	0.6361	0.2348	0.3557	0.1411	1.2379	1.2288
Fe^{3+*}	1.4121	0.5213	0.7896	0.3132	2.7481	2.7279
Cr#	0.771	0.295	0.502	0.273	0.766	0.864
Mg#	0.017	0.675	0.443	0.824	0.025	0.0226
	Cr-magnetite	Cr-spinel	Cr-spinel	Cr-spinel	magnetite	magnetite

**Calculated stoichometrically.*

Table 6.
Chemical composition (wt%) and atom per formula unit (apfu) of spinel group on the basis of 4 O (samples CV-25, CV-41, CV-67).

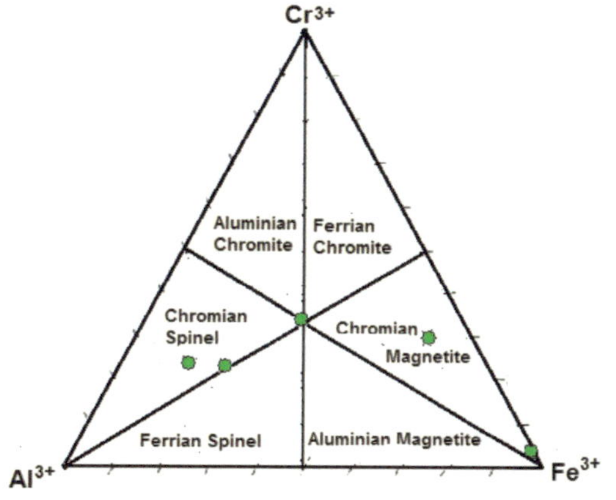

Figure 8.
Cr-Al-Fe plot of Chalki basalt spinels shown as magnetite, chromian magnetite, and chromian spinel (after [14]).

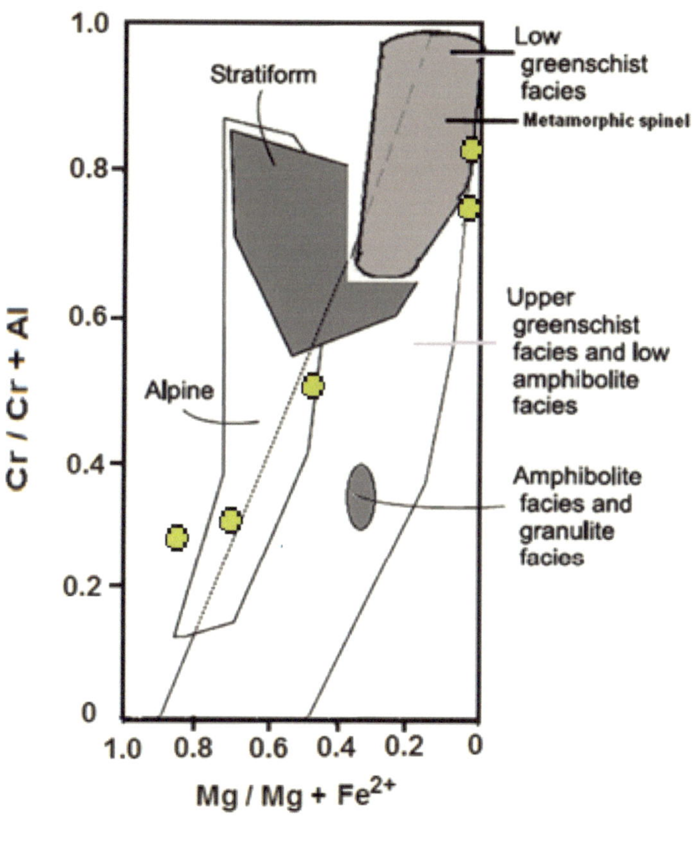

Figure 9.
Cr# vs. Mg# of Chalki basalt spinels indicating their metamorphic origin as a result of low-grade greenschist metamorphism and alpine type (after [15]).

Wt%	1	2	3	4	5	6	7
SiO_2	47.36	46.75	40.64	41.80	41.84	42.52	41.36
TiO_2	0.03	0.00	0.00	0.00	0.06	0.00	0.03
Al_2O_3	0.48	0.63	0.17	0.35	0.86	0.60	0.38
Cr_2O_3	0.01	0.00	0.01	0.02	0.01	0.03	0.00
FeO*	9.88	11.22	10.50	15.14	13.84	13.71	11.09
MnO	0.20	0.10	0.19	0.21	0.25	0.18	0.26
NiO	0.27	0.13	0.23	0.14	0.10	0.19	0.01
ZnO	0.14	0.24	0.58	0.21	0.27	0.00	0.30
MgO	33.21	30.50	36.36	28.77	29.53	29.35	31.96
CaO	0.27	0.42	0.11	0.98	0.32	0.37	0.14
Na_2O	0.07	0.05	0.03	0.03	0.04	0.00	0.03
F	0.01	0.09	0.09	0.09	0.05	0.09	0.09
Cl	0.02	0.02	0.02	0.02	0.02	0.01	0.00
Total	92.0	90.2	88.9	87.8	87.2	87.1	85.6
apfu							
Si	4.298	4.354	3.911	4.141	4.132	4.190	4.107
Ti	0.002	0.000	0.000	0.000	0.004	0.000	0.002
Al	0.052	0.069	0.019	0.041	0.100	0.070	0.044
Cr	0.001	0.000	0.001	0.001	0.001	0.002	0.000
Fe	0.750	0.874	0.845	1.255	1.143	1.130	0.921
Mn	0.016	0.008	0.015	0.018	0.021	0.015	0.022
Ni	0.020	0.010	0.018	0.011	0.008	0.015	0.001
Zn	0.009	0.017	0.041	0.015	0.019	0.000	0.022
Mg	4.493	4.234	5.216	4.249	4.347	4.312	4.732
Ca	0.027	0.042	0.011	0.104	0.034	0.039	0.015
Na	0.012	0.009	0.006	0.005	0.007	0.000	0.005
$\sum$	*9.679*	*9.616*	*10.082*	*9.840*	*9.817*	*9.774*	*9.871*

Table 7.
Chemical composition (wt%) and atom per formula unit (apfu) of serpentines on the basis of 14 O (CV-67).

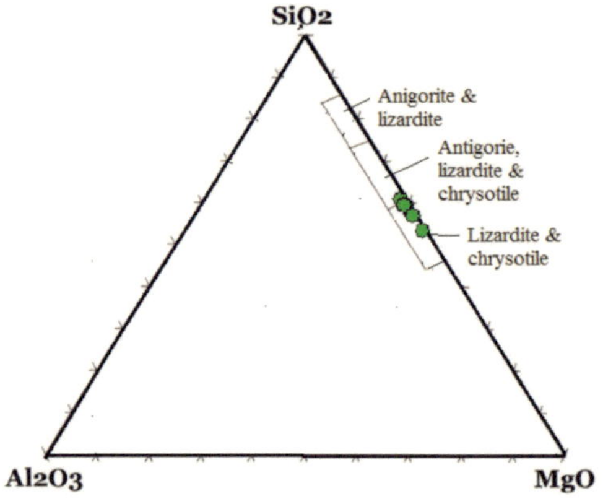

Figure 10.
Chalki basalt serpentines shown as type lizardite and chrysotile (after [16]).

4.2.7 Serpentine

Another less abundant secondary mineral after olivine in Chalki basalt is the serpentine. Their chemistry (**Table 7**) and plot (**Figure 10**) reveal clearly that they are mostly of lizardite and chrysotile types.

5. Petrological significance

1. Olivine alteration to chlorite, serpentine, iddingsite, amphibole, and iron oxide as shown petrographically is a good evidence to low-grade metamorphism.

2. The variation of the plagioclase in Chalki basalt from labradorite (An_{51-61}) to andesine (An_{35-41}) to andesine (An_{22}) reflects the andesitic variety of these volcanic rocks in addition to basalt, while the coexisting albite composition ($An_{0.1\text{-}04}$) indicates clearly that the rocks had suffered albitization due to low-grade sub-sea metamorphism.

3. The NiO content (0.11–0.12 wt%) in the Chalki forsteritic olivine (Fo_{80-81}) reflects the tectonic origin of this extrusion.

4. The enrichment of Chalki pyroxene in MgO causes its shifting from diopside-augite trend of the layered igneous rocks to endiopside (en_{66-68} wo_{27-28} fs_{05-06}) and subcalcic augite (en_{72} wo_{14} fs_{14}).

5. The prevalent chlorite (mainly type diabantite and penninite) in all the Chalki samples represents its suffering from chloritization process after the ferromagnesian olivine and pyroxene.

6. Serpentine (type lizardite and chrysotile) is recorded as an alteration product after the forsteritic olivine, in addition to another rare amphibole secondary mineral of type magnesiohornblende.

7. The spinel group as accessory minerals in Chalki basalt is classified as magnetite, chromian magnetite, and chromian spinel confirming the metamorphic origin (low-temperature sub-sea metamorphism and also of alpine type).

Author details

Mohsin M. Ghazal, Ali I. Al-Juboury* and Sabhan M. Jalal
Geology Department, College of Sciences, Mosul University, Mosul, Iraq

*Address all correspondence to: alialjubory@yahoo.com

References

[1] Dunnington HV, Wetzel R, Morton DM. Mesozoic and Paleozoic. In: van Bellen RC, Dunnington H, Wetzel R, Morton DM, editors. Lexique Stratigraphique International. Paris: Centre National recherché Scientifique Fasc 10a, Iraq; 1959. p. 333

[2] Seilacher A. Kaledonischer Unterbau der Irakiden. Neases Jahrb. Geol. Palent. Abt. A. Monatshefte No 10, Stuttgart. 1963

[3] Buday T. The Regional of Iraq. Stratigraphy and Palaeogeography. Vol. 1. Baghdad, Iraq: State Organization for Minerals; 1980. p. 445

[4] van Bellen RC, Dunnington H, Wetzel R, Morton DM. Lexique Stratigraphique International. Paris: Centre National Recherché Scientifique Fasc 10a, Iraq; 1959. p. 333

[5] Sharland PR, Archer R, Casey DM, Davies RB, Hall SH, Heward AP, et al. Arabian Plate Sequence Stratigraphy. Bahrain: GeoArabia Special Publication 2, Gulf PetroLink; 2001. p. 371

[6] Al-Husseini MI. Pre-Unayzah unconformity, Saudi Arabia. Bahrain: GeoArabia Special Publication 3, Gulf PetroLink; 2004. pp. 15-59

[7] Davoudzadeh M, Weber-Diefenbach K. Contribution to the paleogeography, stratigraphy and tectonics of the upper Paleozoic of Iran. Neues Jahrbuch für Geologie und Paläontologie Abhandlungen. 1987;**175**(2):121-146

[8] Sissakian VK. Geological Map of Iraq. Baghdad: Geological Survey; 2000

[9] Tröger WE. Optische Eigenschaften und Bestimmung der Wichtigsten Gestein-bildenden Minerale, (Feldspata). In: Freund H, editor. Handbuch der Mikroskopie in der Tecknik. Vol. 4. 1st ed. Frankfurt: Umschau Verlag; 1955. pp. 79-119

[10] Leblanc M, Dupuy C, Merlet C. Nickel content of olivine as discriminatory factor between tectonite and cumulate peridotite in ophiolites. Sciences Géologiques. Bulletin. 1984; **37**(2):131-135

[11] Harlbut JCS. Klein C. Manual of Mineralogy (after Dana, J.D.). 19th ed. New York: John Wiley and Sons; 1977. p. 532

[12] Leake BE, Wooley AR, Arps CES, Gilbert MC, Grice JD, Hawthorne FC, et al. Nomenclature of amphiboles: Report of the subcommittee on amphiboles of the international mineralogical association, commission on new minerals and mineral names. The Canadian Mineralogist. 1997;**35**: 219-246

[13] Hey MH. A new revision of the chlorite. Mineralogical Magazine. 1954; **30**:277-292

[14] Steven RE. Composition of some chromites of western hemisphere. American Mineralogist. 1944;**26**:1-34

[15] Irvine TN. Chromian spinel as a petrogenetic indicator, part II. Petrological applications. Canadian Journal of Earth Sciences. 1967;**4**:71-103

[16] Wickes FJ, Plant G. Electron microprobe and X-ray microbeam studies of serpentine textures. Canadian Mineralogist. 1979;**17**:785-830

Chapter 3

Titanite from Titanite-Spots Granodiorites of the Moldanubian Batholith (Central European Variscan Belt)

Miloš René

Abstract

Titanite-rich granodiorites occurring in the Austrian Mühlviertel are intimately associated with the I/S-granites of the Mauthausen/Freistadt granite suite. These rocks form small irregular bodies in granites of this granitic suite of the Moldanubian batholith that are represented by usually fine-grained and dark granodiorites, which contain a large amount of titanite hell "spots" formed by aggregates of plagioclases and quartz. Titanite as a relative plentiful accessory mineral exists in the center of these "spots" as idiomorphic and sphenoidal grains. The composition of titanite ranges from 89 to 92 mol.% titanite end-member. According to its Al concentration, the analyzed titanites could be considered as low-Al titanites (Al = 0.05–0.08 atoms per formula unit). Titanite contains low concentrations of both (Al + Fe^{3+})–OH (2–9 mol.%) and (Al + Fe^{3+})–F (0–8 mol.%). Titanite together with Na-enriched plagioclase and quartz is originated during late-magmatic evolution of titanite-spots granodiorites.

Keywords: titanite, petrology, geochemistry, mineralogy, Bohemian Massif, Moldanubian zone

1. Introduction

In the SW part of the Moldanubian batholith, titanite-rich granitic rocks occur, which were described in the past as titanite-spots granites and diorites. These rocks were described in detail at the Austrian Mühlviertel, SSW of Aigen [1], and at the southern part of the Fürstenstein pluton [2, 3]. For titanite-spots granitic rocks at the Fürstenstein pluton, a local name "engelburgite" was used [2, 4]. In these usually fine-grained and dark granitic rocks, a significant large amount of titanite clots mantled by hell aggregates of feldspars and quartz are found. Although many petrographic studies on these titanite-spots granitic rocks have been performed several decades ago, the origin of these clots remains uncertain [2–11]. The main aim of the present paper is the detailed significance of the petrology and mineralogy of these rocks, in addition to the chemical composition of titanite enclosed in hell "spots".

محتار

2. Geological setting

The titanite-spots granodiorites at the northern Mühlviertel (Austria) occur as small irregular intrusive bodies, concentrated in a local SSW-NNE zone, W of Schlägl. According the most recent geological map 1:200,000 [11], a biggest body that was sampled in the Schlägl quarry forms N-S intrusion, W of Schlägl. These magmatic bodies are enclosed in bigger intrusions of the fine- to medium-grained Mauthausen granites (**Figure 1**). The Mauthausen granites are part of the youngest Freistadt/Mauthausen I/S-type suite (320–310 Ma) of the Moldanubian batholith [12–16]. The titanite-spots granodiorites form small irregular bodies enclosed in these granites.

The titanite-spots granodiorites and tonalites from the Fürstenstein pluton occur in the SSW part of the Bavarian Forest (Germany) forming smaller bodies and blocks enclosed in the medium-grained, porphyritic Saldenburg biotite granites, which form the main part of the Fürstenstein pluton [3, 10, 17]. This pluton represents the largest intrusive body in the Bavarian Forest (**Figure 2**). These granites are equivalent to the Weinsberg medium- to coarse-grained, porphyritic biotite granites at the southern part of the Moldanubian batholith [3, 11]. The titanite-spots granodiorites and tonalites occur predominantly in contact zones between the Saldenburg granites and the medium- to coarse-grained Tittling biotite granites and granodiorites. Their age as determined by Chen and Siebel [10] is 321 ± 4 Ma, by the U/Pb dating method in titanite.

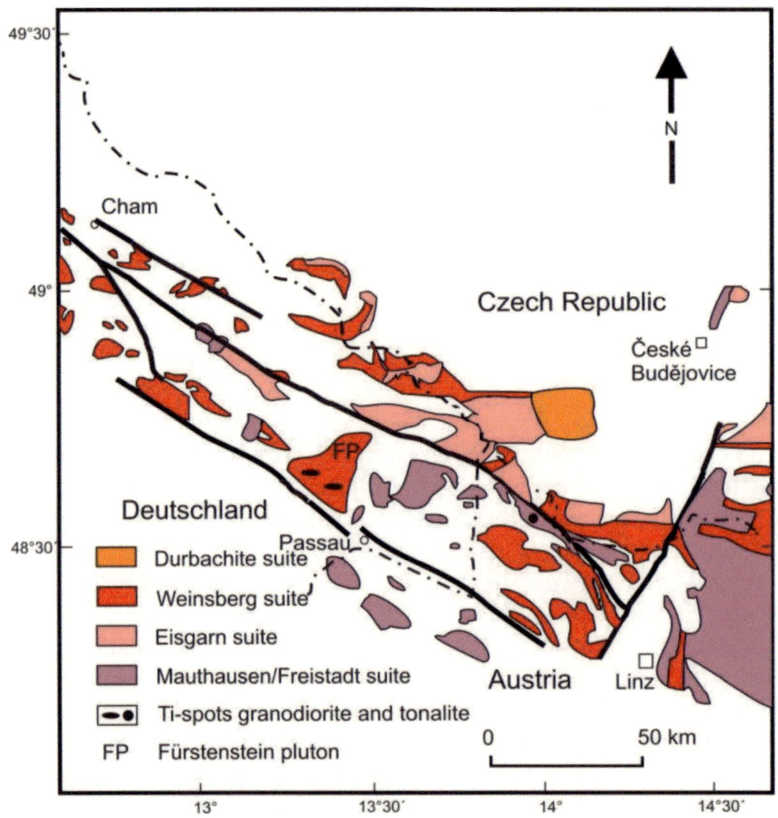

Figure 1.
Geological sketch map of the Šumava Moldanubian batholith branch after [1, 2], modified by the author.

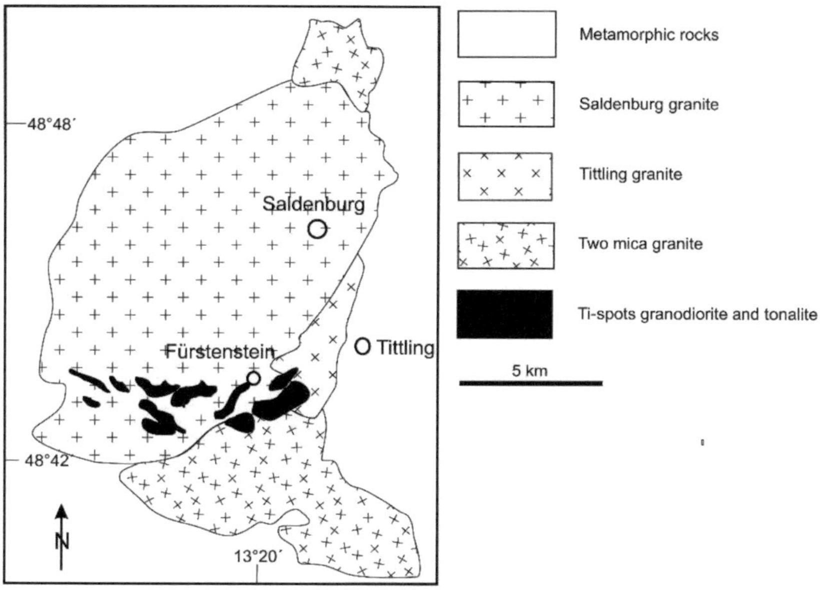

Figure 2.
Geological sketch map of the Fürstenstein pluton after [10], modified by the author.

3. Analytical methods

Approximately 30 quantitative electron probe microanalyses of titanite were performed in the three representative samples of titanite-spots granodiorites from the western Mühlviertel area (Austria). Minerals were analyzed in polished thin sections, and back-scattered electron images (BSE) were acquired to study the interaction of examined titanite with surrounding minerals and the internal structure of individual mineral grains. Element abundances of Al, Ca, Ce, F, Fe, La, Mg, Na, Nb, Nd, Pr, Si, Ta, Th, Ti, Y, and Zr in selected accessory titanite were determined using a CAMECA SX 100 electron probe microanalyzer (EPMA) operated in wavelength-dispersive mode. The contents of the abovementioned elements were determined using an accelerating voltage and beam current of 15 keV and 20 nA, respectively, with a beam diameter of 2–5 μm. The following standards, X-ray lines, and crystals (in parentheses) were used: AlKα, sanidine (TAP); CaKα, fluorapatite (PET); CeLα, $CePO_4$ (PET); FKα, topaz (PC1); FeKα, almandine (LIF); LaLα, $LaPO_4$ (PET); MgKα, Mg_2SiO_4 (TAP); NbLα, columbite (PET); NdLβ, $NdPO_4$ (LIF); PrLβ, $PrPO_4$ (LIF); SiKα, sanidine (TAP); TaLα, $CrTa_2O_6$ (LIF); TbLα, $TbPO_4$ (LIF); ThMα, $CaTh(PO_4)_2$ (PET); TiKα, anatas (PET); YLα, YPO_4 (PET); and ZrLα, zircon (TAP). Intra-REE overlaps were partially resolved using Lα and Lβ lines. The raw data were converted into concentrations using appropriate PAP matrix corrections [18]. The detection limits were approximately 400 ppm for Y and 180–1700 ppm for REE.

4. Results

4.1 Petrography

The titanite-spots granodiorites from the northern Mühlviertel are dark, fine-grained rocks with many light spots formed by plagioclase and bigger, usually idiomorphic titanite crystals (**Figures 3** and **4**).

Figure 3.
Titanite-spots granodiorite from the Schlägl quarry (Mühlviertel, Austria) (photo by René).

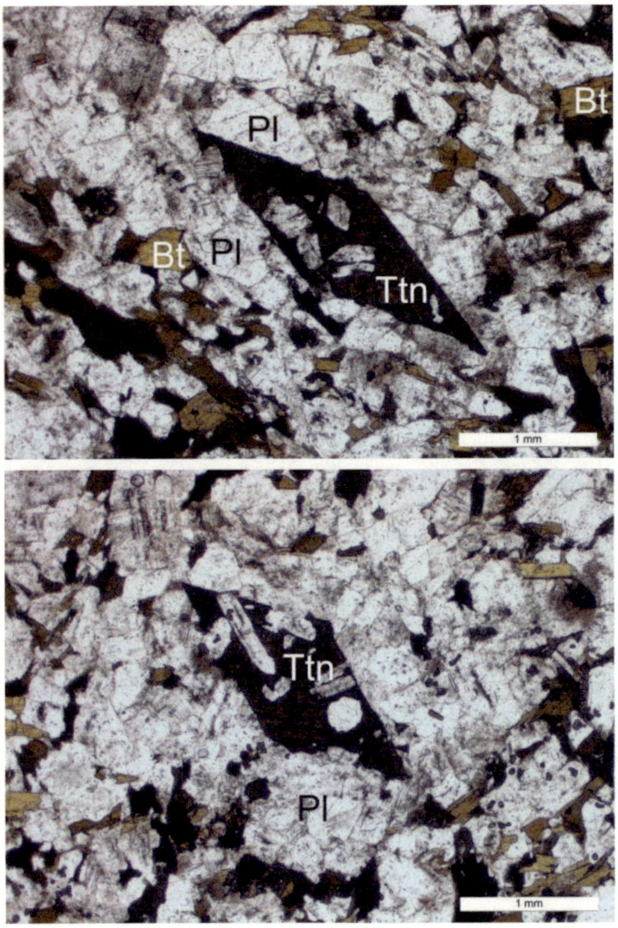

Figure 4.
Microphotography of titanite-spots granodiorite (Ttn, titanite; Pl, plagioclase; Bt, biotite) (photo by René).

The titanite-spots granodiorites contain plagioclase ($An_{30–55}$) (36–46 vol.%), biotite (15–32 vol.%), quartz (21–28 vol.%), K-feldspar (6–16 vol.%), rare amphibole (0–1 vol.%), and pyroxene (0–1 vol.%). Plagioclase grains in light "spots,"

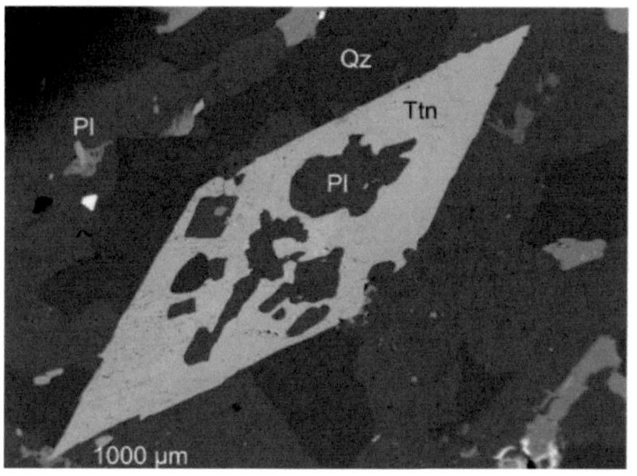

Figure 5.
BSE image of titanite from titanite-spots granodiorite in Mühlviertel area, Austria. Ttn, titanite; Pl, plagioclase; Qz, quartz.

Sample, wt.%	1719–66	1725–15	1725–16	1725–21	1725–23	1725–27
SiO_2	30.43	30.55	30.68	30.18	29.99	30.44
TiO_2	37.20	37.13	36.70	36.91	36.25	37.06
Nb_2O_5	0.30	0.28	0.19	0.28	0.41	0.38
Al_2O_3	1.61	1.30	1.57	1.37	1.55	1.39
Fe_2O_3	1.08	1.33	1.52	1.37	1.57	1.43
CaO	28.51	27.49	28.08	27.28	26.85	27.21
Na_2O	0.01	b.d.l.	0.01	0.02	0.02	0.01
La_2O_3	b.d.l.	0.08	0.02	0.10	0.12	0.15
Ce_2O_3	0.02	0.54	0.36	0.81	0.84	0.86
Pr_2O_3	0.14	0.02	0.16	0.27	0.34	0.11
Nd_2O_3	b.d.l.	0.50	0.28	0.78	0.99	0.90
Y_2O_3	b.d.l.	0.32	0.19	0.37	0.64	0.48
F	0.61	0.33	0.44	0.28	0.25	0.23
O=F	0.26	0.14	0.19	0.12	0.11	0.10
Total	99.65	99.70	100.02	99.90	99.71	100.55
Si	1.00	1.00	1.00	1.00	1.00	1.00
Ti	0.92	0.91	0.90	0.92	0.91	0.92
Nb	0.00	0.00	0.00	0.00	0.06	0.01
Al	0.06	0.05	0.06	0.05	0.06	0.05
Fe^{3+}	0.03	0.03	0.04	0.03	0.04	0.04
Ca	1.04	0.96	0.98	0.97	0.96	0.96
Na	0.00	0.00	0.00	0.00	0.00	0.00
La	0.00	0.00	0.00	0.00	0.00	0.00
Ce	0.00	0.01	0.00	0.01	0.01	0.01
Pr	0.00	0.00	0.00	0.00	0.00	0.00
Nd	0.00	0.01	0.00	0.01	0.01	0.01
Y	0.00	0.01	0.00	0.01	0.01	0.01
F	0.06	0.03	0.05	0.03	0.03	0.02

Sample, wt.%	1719–66	1725–15	1725–16	1725–21	1725–23	1725–27
OH	0.03	0.05	0.05	0.06	0.07	0.07
X(Ttn)	0.91	0.92	0.91	0.91	0.90	0.91
X(Al,Fe^{3+}–F)	0.06	0.03	0.05	0.03	0.03	0.02
X(Al,Fe^{3+}–OH)	0.03	0.05	0.05	0.06	0.07	0.06

b.d.l., below detection limit.

Table 1.
Representative analyses of titanite in titanite-spots granodiorites and tonalites in Mühlviertel area, Austria. Concentration of OH was calculated based on OH´ = (Al + Fe^{3+})–F, according to [19].

which are enriched in albite component ($An_{33–39}$), contain bigger titanite crystals in their cores (**Figures 4** and 5). Biotite is annite with Fe/(Fe + Mg) = 0.51–0.54 and 0.24–0.54 apfu (atoms per formula unit). Amphibole is cummingtonite with Mg/(Mg + Fe) 0.44–0.45 apfu. The accessory minerals are titanite, apatite, zircon, magnetite, ilmenite, allanite, and pyrite.

4.2 Mineralogy and mineral chemistry of titanite

Titanite as a relatively plentiful accessory mineral usually has up to 1–2 mm long. The idiomorphic, sphenoidal crystals are enclosed in light "spots" of fine-grained aggregates of plagioclase and quartz (**Figures 3** and **4**). In some cases, larger titanite crystals contain irregular inclusions of plagioclase (**Figures 4** and 5). Titanite in thin sections is usually dark brown and indistinctly pleochroic. It is compositionally homogeneous, that is, without obvious zoning. The electron microprobe data show that the chemical composition of titanite in these titanite-spots granodiorites and tonalites varies over a wide range. The formula of titanite is calculated on the basis of 1 Si as suggested by Harlov et al. [19] (**Table 1**).

The composition of titanite ranges from 89 to 92 mol.% titanite end-member. The Al and Fe^{3+} contents range from 0.05 to 0.08 atoms per formula unit (apfu) and 0.02–0.05 apfu, respectively. Titanite shows some Al + Fe^{3+} excess over F, which indicates the coupled substitution of (Al + Fe^{3+}) below the line (Al + Fe^{3+}) + F^- = Ti^{4+} + O^{2-} (**Figure 6**). The presence of this substitution in analyzed

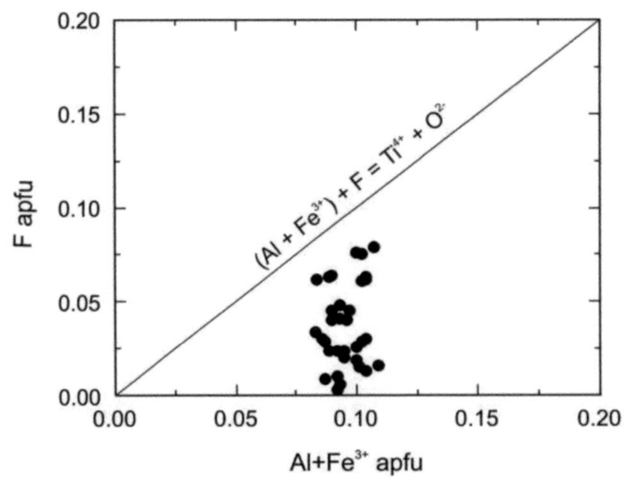

Figure 6.
(Al + Fe^{3+}) vs. F plot indicating additional (Al + Fe^{3+})–OH component in analyzed titanites.

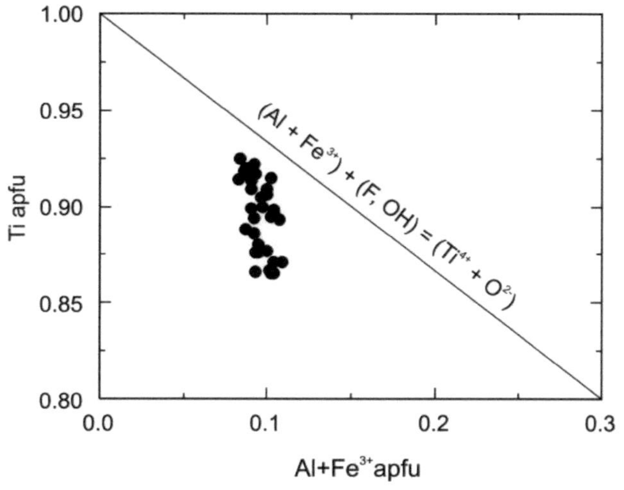

Figure 7.
($Al + Fe^{3+}$) vs. Ti plot indicating additional ($Al + Fe^{3+}$) + (F, OH) = (Ti^{4+} + O^{2-}) substitutions in analyzed titanites.

titanites is shown also in the plot of Ti vs. ($Al + Fe^{3+}$) (**Figure 7**). Calculation of OH content allows the estimation of ($Al + Fe^{3+}$)–OH component from 2 to 9 mol.%. The content of ($Al + Fe^{3+}$)–F component is lower than the amount of ($Al + Fe^{3+}$)–OH component and ranges from 0 to 8 mol.%. The content of REE in analyzed titanite is usually below the microprobe detection limits. However, the contents of La, Ce, Pr, and Nd in some cases could be up to 0.99 wt.% (La_2O_3 up to 0.15 wt.%, Ce_2O_3 up to 0.86 wt.%, Pr_2O_3 up to 0.34 wt.%, Nd_2O_3 up to 0.99 wt.%).

5. Discussion

Titanite, with its considerable variation on chemical composition, is a highly sensitive indicator of oxygen fugacity (fO_2) and water fugacity (fH_2O) during the evolution of geological environment (e.g., [19–25]). The chemical composition of the analyzed titanite shows that the substitution (Al, Fe^{3+}) + F = Ti^{4+} + O^{2-} and the substitution (Al, Fe^{3+}) + OH = Ti^{4+} + O^{2-} were the most significant substitutions in these titanites (**Figures 6** and 7). In both substitutions, it could be up to 30% of Ti replaced by aluminum and ferric iron [21]. According to their F, Al, and Fe^{3+} concentrations, the titanites considered as low-Al titanites could be analyzed [26]. Their low F and Al content could be well compared with contents of both elements in similar magmatic titanites [27, 28]. The partial Al + Fe^{3+} excess over F indicates the presence of Al-involving substitution reaction such as (Al, Fe^{3+}) + OH = Ti + O (**Figure 4**).

The origin of titanite that is usually concentrated in light spots is thought to be formed with albite-enriched component plagioclases (oligoclase) and quartz on the expense of andesine and Ti-annite as expressed by the following reaction:

$$\text{Andesine} + \text{Ti} - \text{rich annite} = \text{Titanite} + \text{Oligoclase} + \text{Quartz} \qquad (1)$$

Similar late-magmatic reaction is supposed for titanite origin from I-type granites occurred in the Tříbeč Mountains (Western Carpathians, Slovakia) [28]. This suggestion for the origin of titanite-spots granodiorites in the Austrian Mühlviertel by late-magmatic crystallization was presented in the past by Rupp et al. [11]. Late-magmatic origin of titanite in titanite-spots granodiorites and tonalites from the Fürstenstein pluton is supposed by Chen and Siebel [10]. In the past, the origin of these titanite-spots granitic rocks in the Fürstenstein pluton was given in detailed discussion of older papers [6–8] and also supposed by Troll [3]. Especially according to old detailed studies of titanite-spots granodiorites and tonalites from the Fürstenstein pluton, Drescher [6], Fischer [7], and Troll [3] supposed that titanite in the light spots originated by late-magmatic and/or younger hydrothermal crystallization. In all these cases, biotite served as a source of Ti for titanite, and titanite is most likely originated from a reaction involving biotite and probably anorthite component in plagioclase, during its deanorthization.

Incorporation of the REE in titanite is realized by the coupled substitution:

$$Ca^{2+} + Ti^{4+} = REE^{3+} + (Al, Fe)^{3+} \quad (2)$$

Partial enrichment of REE, especially Ce and La, is significant for magmatic titanite, especially for titanite occurred in alkalic magmatic rocks and lamprophyres [29, 30]. Incorporation of Nb and Ta is realized by the following reaction:

$$2Ti^{4+} = (Nb, Ta)^{5+} + (Al, Fe)^{3+} \quad (3)$$

Both the exchange reactions (2 and 3) are significant for primary magmatic titanite [21]. The Nb and Ta enriched titanite was described from some granitic pegmatites [31].

6. Conclusions

Titanite-rich granodiorites occur in the Austrian Mühlviertel as small irregular bodies enclosed in I/S-type granites of the Freistadt/Mauthausen suite, which are usually fine-grained and dark granitic rocks having significant large amount of titanite spots mantled by hell aggregates of plagioclase and quartz.

Titanite as a relative plentiful accessory mineral usually forms idiomorphic, sphenoidal grains enclosed in light "spots" of fine-grained aggregates of plagioclases and quartz. The composition of titanite ranges from 89 to 92 mol.% titanite end-member. According to its Al concentration, the analyzed titanites could be considered as low-Al titanite with an aluminum of 0.05–0.08 apfu. The Fe^{3+} contents range from 0.02 to 0.05 apfu. Titanite shows some Al + Fe^{3+} excess over F, which indicates the presence of coupled substitution of (Al + Fe^{3+}) + (F, OH) = Ti^{4+} + O^{2-}. Calculation of OH content allows the estimation of (Al + Fe^{3+})–OH component from 2 to 9 mol.%. The content of (Al + Fe^{3+})–F component is lower than the amount of (Al + Fe^{3+})–OH component and ranges from 0 to 8 mol.%. Titanite of these titanite-spots granodiorites together with plagioclase and quartz was originated during late-magmatic evolution. The origin of titanite can be described by the following reaction:

$$\text{Andesine} + \text{Ti} - \text{rich annite} = \text{Titanite} + \text{Oligoclase} + \text{Quartz} \quad (4)$$

Acknowledgements

Thanks are extended to the support of the Long-Term Conceptual Development Research Organization RVO 67985891 and the project of Ministry of Education, Youth and Sports (ME10083). I am also grateful to P. Gadas and R. Škoda from Institute of Masaryk University for their technical assistance by using electron microprobe analyses of selected minerals (titanite, plagioclase, biotite, amphibole).

I am also grateful to the academic editor Ali Al-Juboury for the numerous comments and recommendations that helped to improve this paper.

Conflict of interest

The author declares no conflict of interests.

Author details

Miloš René
Institute of Rock Structure and Mechanics, Academy of the Sciences of the Czech Republic, Prague, Czech Republic

*Address all correspondence to: rene@irsm.cas.cz

References

[1] Fuchs G, Thiele O. Erläuterungen zur Übersichtskarte des Kristallins in westlichen Mühlviertel und im Sauwald, Oberösterreich. Geologische Bundesanstalt: Wien; 1968. p. 96

[2] Frenzel A. Das Passauer Granitmassiv. Geognostische Jahreshefte. 1911;**24**:105-192

[3] Troll G. Das Intrusivgebiet von Fürstenstein. Geologica Bavarica. 1964;**52**:1-140

[4] Graber HV. Die Diorite des Passauer Waldes. Geologische Rundschau. 1933;**24**:15-27

[5] Osann A. Über Titanitfleckengranite. Neues Jahrbuch für Mineralogie (Abhandlungen). 1923;**48**:223-239

[6] Drescher RK. Zur Tektonik und Petrographie der Diorite von Fürstenstein. Abhandlungen der Hessischen Geologischen Landes-anstalt. 1925;**8**:1-49

[7] Fischer G. Über Verbreitung und Entstehung der Titanitfleckengesteine im Bayerischen Wald. Zentrbl. f. Min. 1926;**25**:155-168

[8] Scholtz H. Das Saldenburger Granitmassiv und seine Umgebung. Geological Survey (Palaeontology Monographs). 1927;**2**:1-157

[9] Köhler A. Zur Entstehung der Granite der südböhmischen Masse. Mineralogy and Petrology. 1948;**1**:175-184

[10] Chen F, Siebel W. Zircon an titanite geochronology of the Fürstenstein granite massif, Bavarian Forest, NW Bohemian Massif: Pulses of the late Variscan magmatic activity. European Journal of Mineralogy. 2004;**16**:777-788. DOI: 10.1127/0935-1221/2004/0016-0777

[11] Rupp C, Linner M, Mand GW. Geologische Karte von Oberösterreich 1:200000. Erläuterungen. Geologische Bundesanstalt: Wien; 2011. p. 255

[12] Vellmer C, Wedepohl KH. Geochemical characterization and origin of granitoids from the South Bohemian Batholith in Lower Austria. Contributions to Mineralogy and Petrology. 1994;**118**:13-32. DOI: 10.1007/BF00310608

[13] Gerdes A, Wörner G, Henk A. Post-collisional granite generation and HT/LP metamorphism by radiogenic heating: The Variscan South Bohemian Batholith. Journal of the Geological Society of London. 2000;**157**:577-587. DOI: 10.1144/jgs.157.3.577

[14] Gerdes A, Friedl G, Parrish RR, Finger F. High-resolution geochronology of the Variscan granite emplacement—The South Bohemian Batholith. Journal of the Czech Geological Society. 2003;**48**:53-54

[15] Finger F, Gerdes A, René M, Riegler G. The Saxo-Danubian granite belt: Magmatic response to post-collisional delamination of mantle lithosphere below the south-western sector of the Bohemian Massif (Variscan orogeny). Geologica Carpathica. 2009;**60**:205-212. DOI: 10.2478/v10096-009-0014-3

[16] Verner K, René M, Žák J, Janoušek V. A brief introduction in the geology of the Moldanubian batholith. In: Janoušek V, Žák J, editors. Eurogranites 2015: Variscan Plutons of the Bohemian Massif. Post-conference Field Trip Following the 26th IUGG General Assembly in Prague. Prague: Czech Geological Survey; 2015. pp. 103-109

[17] Franzen Ch. Zur Petrologie und Geochemie des Fürstenstein-Intrusivkomplexes (Bayerischer Wald)

[thesis]. Heidelberg: Ruprecht-Karls-Universität; 1999. p. 154

[18] Pouchou JL, Pichoir F. "PAP" (φ-ρ–Z) procedure for improved quantitative microanalysis. In: Armstrong JT, editor. Microbeam Analysis. San Francisco: San Francisco Press; 1985. pp. 104-106

[19] Harlov D, Tropper P, Seifert W, Nijland T, Förster HJ. Formation of Al-rich titanite ($CaTiSiO_4O$–Ca $AlSiO_4OH$) reaction rims on ilmenite in metamorphic rocks as a function of fH_2O and fO_2. Lithos. 2006;**88**:72-84. DOI: 10.1006/J.Lithos.2005.08.00

[20] Wones DR. Significance of the assemblage titanite + magnetite + quartz in granitic rocks. American Mineralogist. 1996;**74**:744-749

[21] Piccoli P, Candela P, Rivers M. Interpreting magmatic processes from accessory phases: Titanite—A small-scale recorder of large-scale processes. Transactions of the Royal Society of Edinburgh: Earth Sciences. 2000;**91**:257-267. DOI: 10.1017/s026359330007422

[22] Xirouchakis D, Lindsley DH, Andersen DJ. Assemblages with titanite ($CaTiOSiO_4$), Ca-Mg-Fe olivine and pyroxenes, Fe-Mg-Ti oxides, and quartz: Part I. Theory. American Mineralogist. 2001;**86**:247-253

[23] Xirouchakis D, Lindsley DH, Frost BR. Assemblages with titanite ($CaTiOSiO_4$), Ca-Mg-Fe olivine and pyroxenes, Fe-Mg-Ti oxides, and quartz: Part II. Application. American Mineralogist. 2001;**86**:254-264

[24] Troitzsch U, Ellis DJ. Thermodynamic properties and stability of AlF-bearing titanite $CaTiSiO_4$–$CaAlFSiO_4$. Contributions to Mineralogy and Petrology. 2002;**142**:543-563. DOI: 10.1007/s004100100309

[25] Tropper P, Manning CE, Essene EJ. The substitution of Al and Fe in titanite in high pressure and temperature: Experimental constraints on phase relations and solid solution properties. Journal of Petrology. 2002;**43**:1787-1814. DOI: 10.1093/petrology/43.10.1787

[26] Oberti R, Smith DC, Rossi G, Caucia F The crystal chemistry of high-aluminium titanites. European Journal of Mineralogy. 1991;**3**:777-792. DOI: 10.1127/ejm/3/5/0777

[27] Enami M, Suzuki K, Liou JG, Bird DK. Al–Fe^{3+} and F-OH substitutions in titanite and constraints on their P–T dependence. European Journal of Mineralogy. 1993;**5**:219-231. DOI: 10.1127/ejm/5/2/0219

[28] Broska I, Harlov D, Tropper P, Siman P. Formation of magmatic titanite and ilmenite–titanite phase relations during granite alteration in the Tříbeč Mountains, Western Carpathians, Slovakia. Lithos. 2007;**95**:58-71. DOI: 10.1016/j.lithos.2006.07.012

[29] Fleischer M. Relation of relative concentrations of lanthanides in titanite to type of host rocks. American Mineralogist. 1978;**63**:869-873

[30] Seifert W. REE-, Zr-, and Th-rich titanite and associated accessory minerals from a kersantite in the Frankenwal, Germany. Mineralogy and Petrology. 2005;**84**:129-146. DOI: 10.1007/s00710-005-0076-6

[31] Russel JK, Groat LA, Halleran AD. LREE-rich niobian titanite from Mount Bisson, British Columbia: Chemistry and exchange mechanisms. The Canadian Mineralogist. 1994;**32**:575-587

Chapter 4

Investigation of the Usability of Pseudoleucites in Central Anatolia Alkali Syenites as Industrial Raw Materials

Zeynel Başibüyük and Gökhan Ekincioğlu

Abstract

Pseudoleucite syenite is a magmatic rock, which is rarely found in the foidolite rock group. With respect to the compositions of similar alkali feldspars as sodium potassium aluminosilicates, feldspathoid minerals are normally characterized by silica deficiency. Pseudoleucite syenite formed from alkaline (sodium and potassium)-rich and silica-poor magmas. In this study, intrusion-related distributions, mineralogical and petrographical properties, and mineral chemistry of pseudoleucites in İsahocalı (Kırşehir) alkali syenites from Central Anatolia Granitoids have been investigated, and magnetic enrichment processes have been carried out on their crushed and grinded samples. As a result of the enrichment of pseudoleucite syenites with a high amount of K_2O + Na_2O (12.25 + 5.61 wt.%), via dry magnetic separator, the obtained data demonstrated that pseudoleucites in İsahocalı Alkali syenites can be used as industrial raw material in sectors such as ceramics, agriculture, cement industries, etc.

Keywords: usability of pseudoleucite, industrial raw material, ceramics, İsahocalı alkali syenites, Kırşehir, Central Anatolia

1. Introduction

Leucite is a rock-forming mineral composed of potassium and aluminum tectosilicate $K[AlSi_2O_6]$. Its crystals have the form of tetragonal but it is isotropic, because of its pseudocubic system [1]. Pseudoleucites are intergrowths of two or more minerals: leucite, nepheline, and K-feldspar; nepheline and feldspar; or analcite, nepheline, and feldspar; or alteration products of these, thought to be pseudomorphing leucite in igneous rocks [2].

In the Central Anatolia Granitoids, studied pseudoleucite syenites, which are feldspathoid igneous rocks, appear as light colored, coarse crystalline, and compositionally silica-poor. They can also be classified as foid syenite, consisting of pseudoleucite (leucite + nepheline + orthoclase) and aegerine-augite minerals.

Recent studies have shown that foid syenitic rocks containing feldspathoid minerals can be used as starting materials for the production of glass, ceramics, and paint. The lack of free silica, high alkaline and alumina content, high melting power, and narrow melting range of foid syenites are the properties that are desired in the glass industry. In addition, they provide high resistance to weather conditions, as well as the use of roof particles, road materials, stone pavements, as well as concrete aggregate and asphalt production. Other potential use areas are artificial fertilizer, refractory, cement mortar, and paper [3, 4].

The İsahocalı (Kırşehir) area in Central Anatolia includes pseudoleucite syenites. Such alkaline syenite intrusions in the region, which is the main objective of this study, offer irregularly shaped outcrops reaching up to 1 km^2. In the vicinity of the study area, there are also nephelite syenites, which are produced by B & S Invest CO. B & S Invest CO, via magnetic enrichment operations. After the magnetic enrichment, the iron-containing minerals are removed and the sodium and potassium-rich industrial raw materials are marketed to the ceramic industry.

The aim of this study is to investigate the usability of sodium- and potassium-rich pseudoleucites in İsahocalı (Kırşehir) alkali syenites from the Central Anatolia Granitoids.

2. Materials and methods

In this study, the spread of pseudoleucite syenites in the İsahocalı (Kırşehir) area has been mapped. Then samples were taken from these intrusions for laboratory studies. In order to determine the mineralogical and petrographic properties of the pseudoleucite syenites samples obtained from the field, thin sections were prepared in the Thin Section Laboratory of the Department of Geological Engineering of the Engineering Faculty of the University of Ahi Evran.

Pseudoleucite syenite samples were grinded by a crusher and ball mill at B&S Yatırım AŞ laboratory. As the enrichment method, magnetic enrichment was carried out to remove the iron-bearing minerals in the grinded samples. The chemical analyses of the samples were carried out by the XRF-method before and after enrichment studies. As a result of chemical analysis, their sodium, potassium, and iron values were compared.

3. Geology

The Central Anatolia, in which the study area is located, consists of several continental blocks separated from each other by ophiolitic suture zones (**Figure 2**). These tectonic belts are, from the north to the south, Pontid Continent, Intra-Pontide suture, Sakarya Continent, Ankara-Yozgat-Erzincan suture (Central Anatolian Metamorphites), and the Kırşehir Continent (Central Anatolian Metamorphics) (**Figure 1**), and the continental blocks developed as a result of Pan-African, Hercynian, and Cimmerian orogenies and remained as the continental basement during the Neo-Tethyan evolution [5]. The Neo-Tethys Ocean was opened by rifting of these continental bases along two lines in Lias, so that the Intra-Pontid and Ankara-Yozgat-Erzincan Ocean branches developed [6]. At the beginning of the Late Cretaceous, a subduction started along the entire Pontide belt [6]; in other words, the northern branch of Neo-Tethys began to subduct under the Pontide. During this period, the greatest ophiolitic nappes including Refahiye Complex, which is the oldest unit of the study area, was settled on Anatolide-Tauride continent [6, 7]. In the Paleocene-Early Eocene, the northern branch of the Neo-Tethys was completely closed and the continent-continent collision [6] took place [8].

The oldest level in the study area consists of Paleozoic Middle Anatolian Metamorphics consisting of schist, gneiss, amphibolite, and marbles. Seymen (1981)

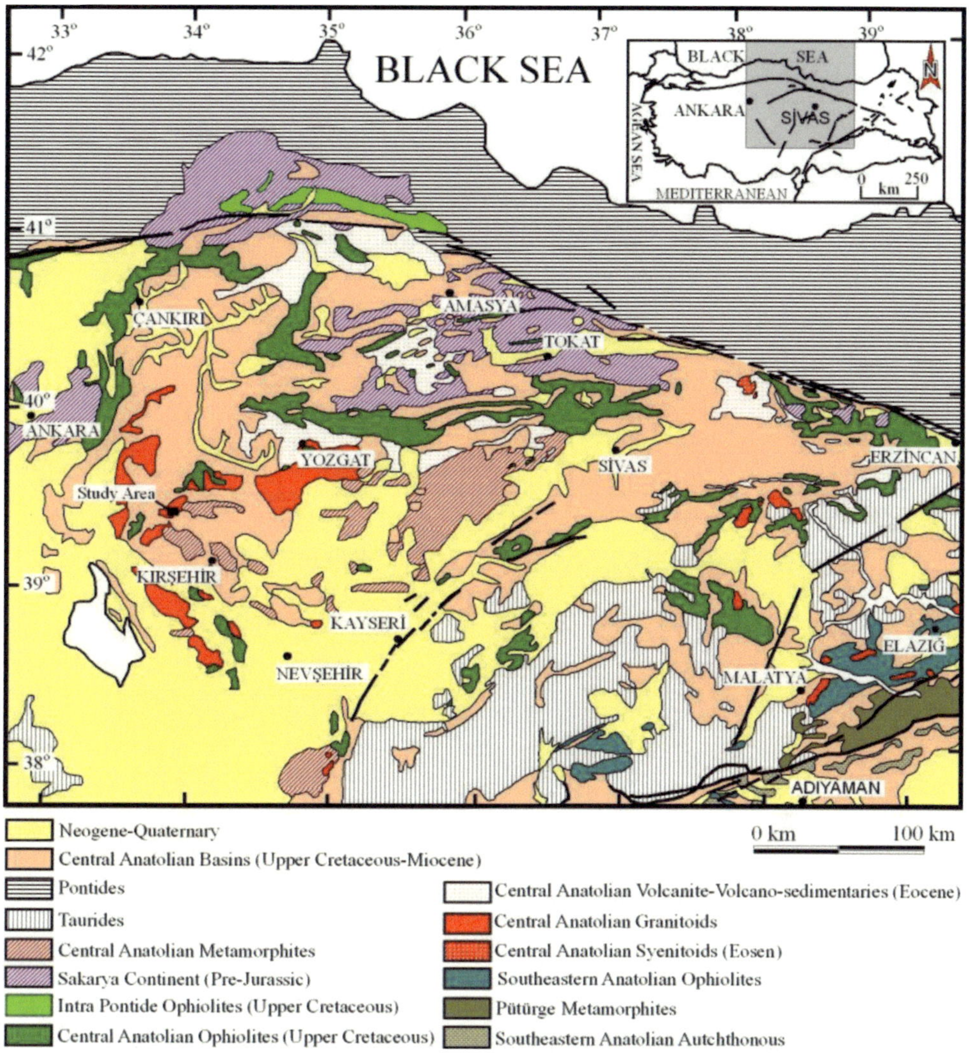

Figure 1.
The regional geological map of the study area [7–11].

divided the unit into three formations in his study [12]. These are from the old to the young, Kalkanlıdağ, Tamadağ, and Bozçaldağ formations. In the Upper Cretaceous, the Central Anatolian Volcanites with dacite-rhyolite-rhyodacite composition and the Upper Cretaceous-Paleocene granite-syenite-monzonite composition of the Central Anatolian Granitoids were cut by coeval units. All these units are unconformably overlain by Lower-Middle Eocene-Quaternary sedimentary units (**Figure 2**).

The igneous rocks in the region and its vicinity are defined as Central Anatolian Granitoids by Erler and Bayhan (1995) [13]. These rocks with granite, syenite, and monzonite composition are Upper Cretaceous-Paleocene aged [14, 15]. Central Anatolian Metamorphics and Central Anatolian Granitoids are covered by Eocene—Quaternary aged, marine, and terrestrial rock units. These units consist of alternation of conglomerate, sandstone, siltstone, claystone, limestone, and occasionally tuff and gypsiferous levels, from old to young, formations defined as Baraklı Formation, Arzilar Limestone Member, Meşeköy Formation, Kozaklı Limestone Member, Kızılırmak Formation, travertine, and alluvium [15–17].

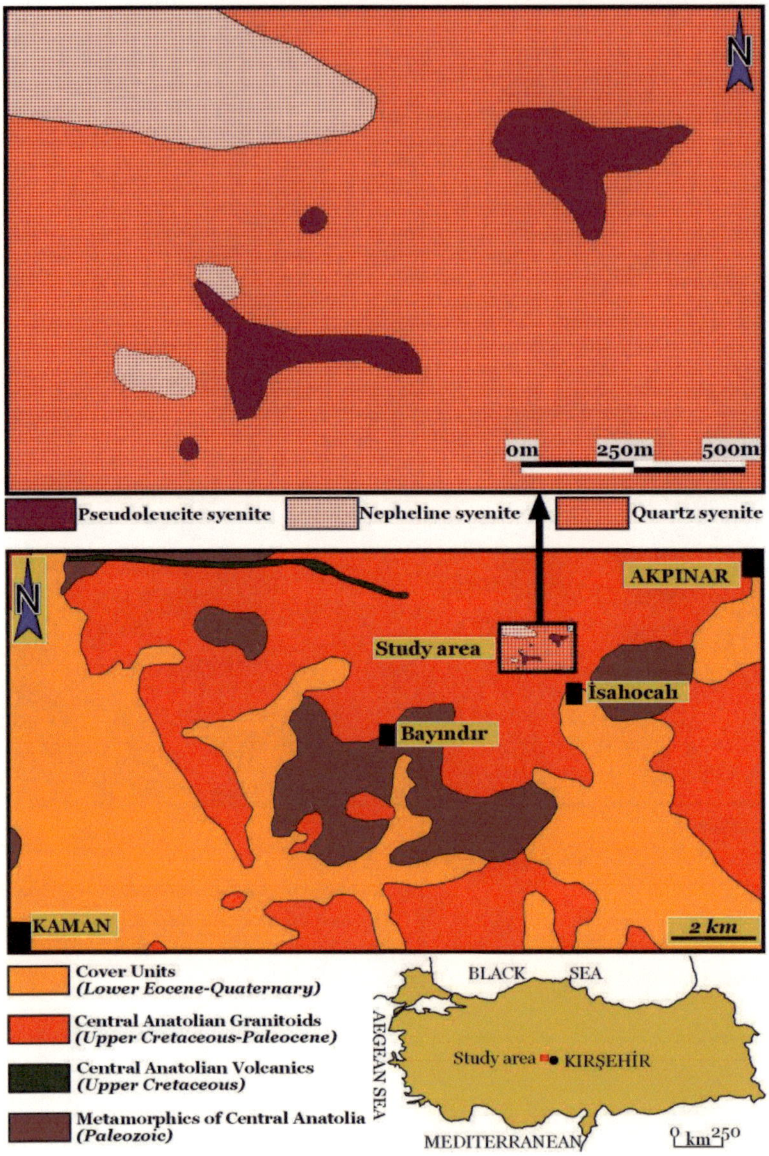

Figure 2.
Geological map of the study area [16].

4. Findings

4.1 Field studies

İsahocalı (Kırşehir) pseudoleucite syenites [18] in Central Anatolian Granitoids are spread along a line extending in the NE–SW direction in the region. They are topographically exposed at the high level of the area. The units have a hard, compact, and massive structure and do not contain alteration. Two dominant colors of intrusions are distinct in the region. At the southwest (SW) end of the study area, they appear as darker color due to the enrichment in the mafic mineral contents, whereas in other locations they have lighter color due to the decrease in mafic mineral content (**Figure 3**). The cooling cracks are also observed in the unit (**Figure 4**).

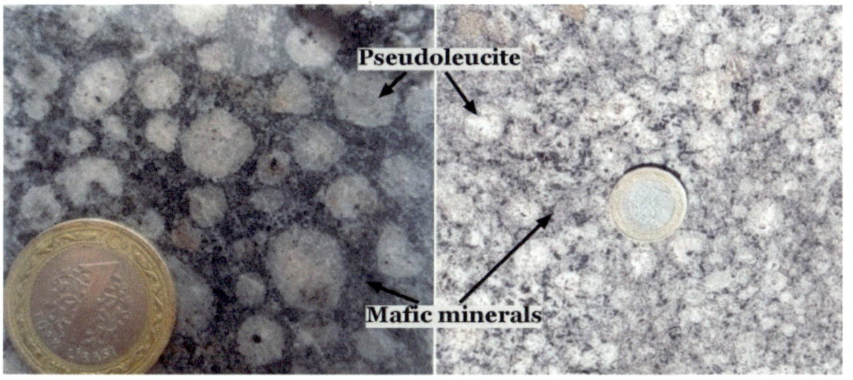

Figure 3.
Pseudoleucite syenite different color selections observed depending on the mafic mineral contents.

Figure 4.
Cooling cracks observed in pseudoleucite syenites.

4.2 Laboratory works

Within the scope of laboratory studies, İsahocalı (Kırşehir) pseudoleucite syenite samples (**Figure 5**) have been processed in the Materials and methods section. The data obtained are given in mineralogical petrographic investigations and ore preparation and enrichment processes.

4.2.1 Mineralogical petrographic investigations

The leucite mineral is derived from magmas with high potassium and low silica contents. It is a typical magmatic mineral, which solidifies as late stage crystals. On the other hand, pseudoleucite generally coexists with potassium feldspar, nepheline, and small amounts of sodalite, kankrinite, and/or zeolite. At the same time, pseudoleucite is defined as a collection of crystals in both volcanic and plutonic rocks showing a leucite crystal structure [19, 20]. As a result of thin section studies, the rock with holocrystalline hypidiomorphic granular texture in the study area, pseudoleucite (leucite + nepheline + orthoclase), aegirine, aegirinaugite, and opaque minerals were determined (**Figures 6** and 7).

At the same time, opacification-type alterations were observed in mafic minerals. According to its mineralogical contents, rock is named foid syenite, which is defined in foidolite rock group from feldspathoid rocks (**Figure 8**).

Figure 5.
View of pseudoleucite samples taken from land.

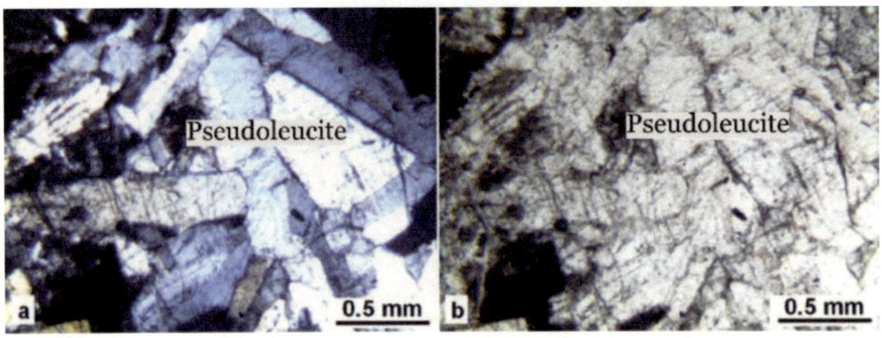

Figure 6.
Leucious prismatic leucite, nepheline, and orthoclase minerals in pseudoleucite.

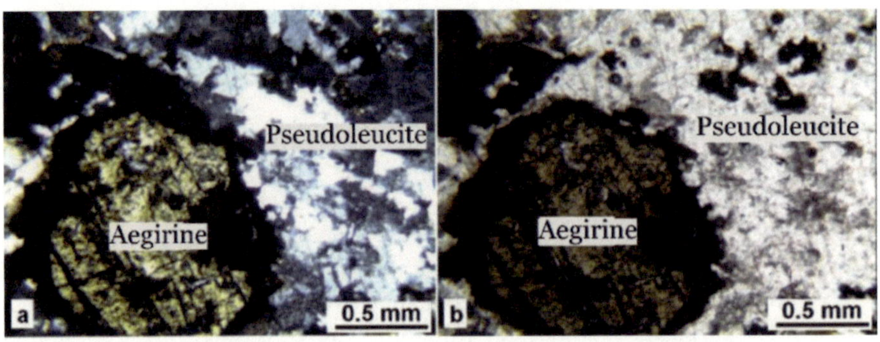

Figure 7.
Euhedral aegirine mineral in pseudoleucite syenite rock.

Similarly, according to the results of chemical analysis (**Table 1**), the rock is classified in the total alkali silica (TAS) diagram. According to the TAS diagram, the analyzed sample with 56.52 wt.% SiO_2 and 15.71 wt.% Na_2O + K_2O contents is plotted in the field of foid syenites (**Figure 9**).

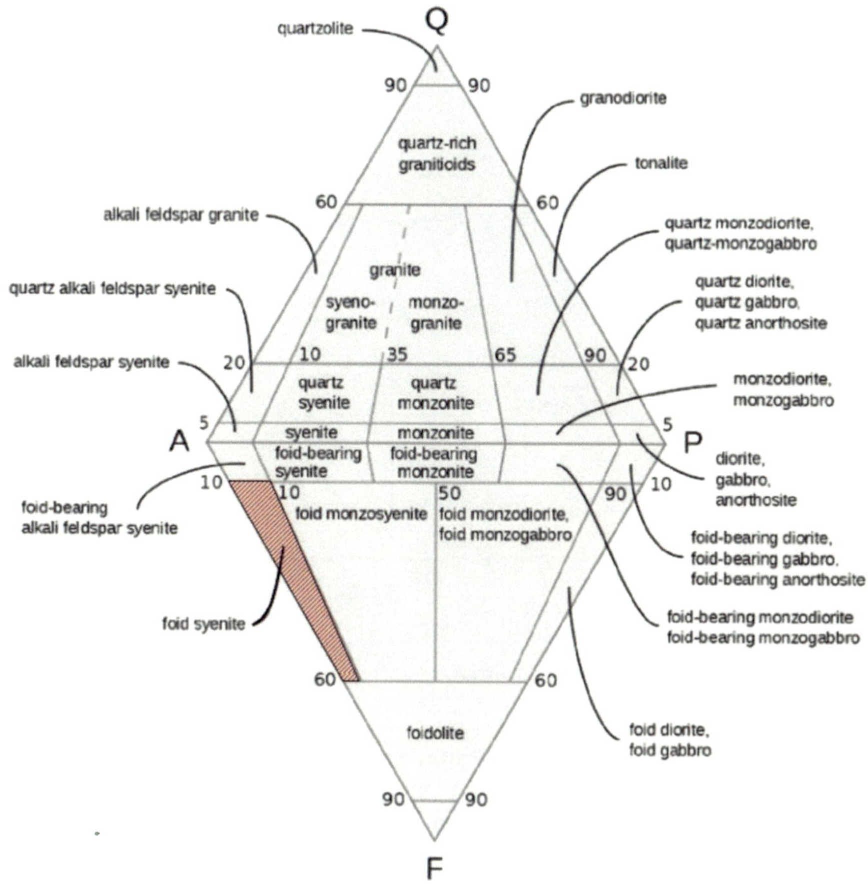

Figure 8.
Modal QAPF classification of magmatic rocks [21].

Chemical content (wt.%)		
	Before enrichment %	**After enrichment %**
SiO_2	56.52	57.54
Al_2O_3	21.01	21.56
K_2O	11.13	12.24
Fe_2O_3	3.11	0.43
CaO	2.02	1.03
Na_2O	4.58	5.61
MgO	0.26	0.04
TiO_2	0.21	0.09
MnO	0.004	0.001
LOI	1.12	1.27

Table 1.
Results of chemical analysis (wt.%) of studied samples.

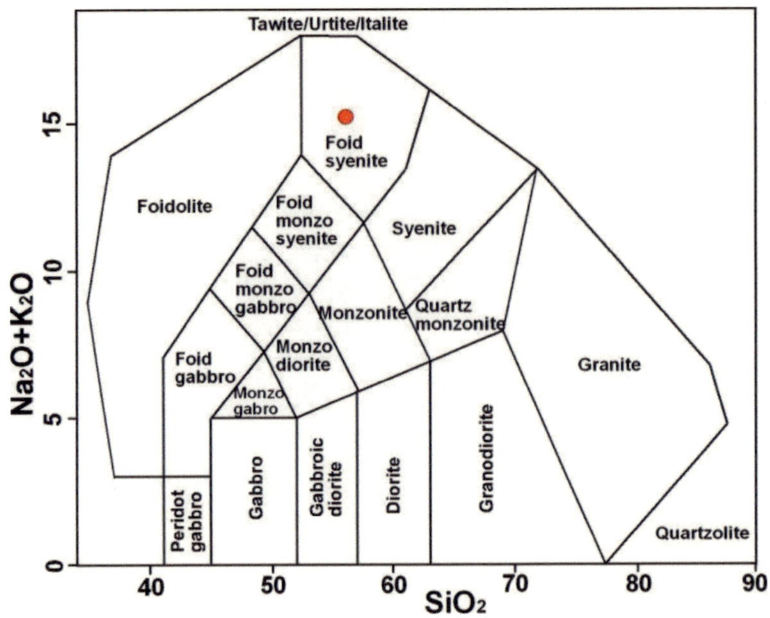

Figure 9.
TAS diagram of analyzed sample in magmatic rock classification [23].

Bayhan conducted investigations around the study area and stated that the mineralogical composition of the alkaline igneous rocks in the region was formed by orthoclase, plagioclase, leucite, nozean, nepheline, cancrinite, aegirinaugite, amphibole, and biotite minerals [22]. We concluded that converting of leucite, nepheline, and orthoclase minerals form euhedral pseudoleucite crystals in these intrusions.

4.2.2 Ore preparation and enrichment processes

4.2.2.1 Sample preparation

Within the scope of this study, approximately 40 kg of sample was provided to represent the study area. Pseudoleucite syenite samples brought to the laboratory were reduced to less than −1 mm size using jaw and hammer crushers. The appearance of the obtained samples after size reduction operations is given in **Figure 10**.

After crushing operations, sample reduction operations were carried out. Then, magnetic enrichment experiments were performed. The appearance of the product obtained after magnetic enrichment is given in **Figure 11**.

4.2.2.2 Magnetic enrichment

Magnetic separation experiments were carried out by Aksa Magnet magnetic separator with 9000 Gauss. The enrichment process was carried out under dry conditions. Magnetic separation was carried out once in the magnetic separator with 5000 gr sample. Then, 2470 gr pseudoleucite concentration was obtained. Chemical analysis of these samples was performed by the Oxford Instruments X-Supreme brand XRF device before and after magnetic separation processes. The results of the analysis are given in **Table 1**.

Figure 10.
The appearance of the obtained samples after size reduction operations.

Figure 11.
The appearance of the product obtained after magnetic enrichment.

5. Results

In this study, the usability of pseudoleucites in alkali syenites from İsahocali region (Kırşehir) in Central Anatolia as a raw material in glass and ceramics sectors has been investigated. Within the scope of the study, approximately 40 kg of sample was provided to represent the studied intrusions. Pseudoleucite syenite samples were reduced to less than −1 mm size using jaw and hammer crushers. After crushing and grinding processes, magnetic enrichment experiments were performed. For this purpose, the samples were subjected to magnetic separation process in the dry magnetic separator. $Na_2O + K_2O$ (15.71 wt.%) and Fe_2O_3 (3.11 wt.%) contents were determined in pseudoleucite syenites prior to the magnetic enrichment process. In terms of the analytical result performed after enrichment processes, their $Na_2O + K_2O$ contents increased (17.76 wt.%) and Fe_2O_3 content (0.43 wt.%) decreased. It was observed that the iron-containing minerals could be removed after the separation process.

We claimed that high concentrations of iron minerals in İsahocalı (Kırşehir) pseudoleucite syenites can easily remove via magnetic enrichment processes under optimum enrichment conditions. It is suggested that the concentrated $Na_2O + K_2O$-rich product obtained after magnetic enrichment can be added to the economy as a high quality product and can be used in different sectors of industry with the necessary tests and investigations. The nepheline syenites produced in the Kırşehir region are used in cement and ceramic industry [23, 24]. More detailed tests on the availability of pseudoleucite syenites in the glass and ceramics industry are recommended.

Acknowledgements

The authors would like to thank BS Investment Inc. for their support in carrying out this study.

Conflict of interest

The authors declare no conflict of interest.

Author details

Zeynel Başibüyük[1] and Gökhan Ekincioğlu[2*]

1 Department of Geological Engineering, Faculty of Engineering and Architecture, Kırşehir Ahi Evran University, Kirşehir, Turkey

2 Department of Mining and Mineral Extraction, Kaman Vocational School, Kırşehir Ahi Evran University, Kirşehir, Turkey

*Address all correspondence to: gokhanekincioglu@gmail.com

References

[1] Karakaya Çelik M, Mineraloji Ö. Bizim Büro; 1993. p. 276 (in Turkish)

[2] Hesselbo SP. Pseudoleucite from the Gardar of South Greenland. Bulletin of the Geological Society of Denmark. 1986;**35**:11-17

[3] VT ML. Nepheline Syenite. In: Kogel JE, Trivedi NC, Barker JM, Krukowski ST, editors. Industrial Minerals & Rocks. 7th ed. Colorado: Published by Society for Mining, Metallurgy, and Exploration, Inc.; 2006. pp. 653-670

[4] Haner S, Demir M. Nefelinli Siyenit: Bir Gözden Geçirme. Jeoloji Mühendisliği Dergisi. 2018;**42**(1): 107-120 (in Turkish)

[5] Tüysüz O. Karadeniz'den orta anadolu'ya bir jeotravers: Kuzey neotetisin tektonik evrimi. Türkiye Petrol Jeolojları Derneği Bülteni. 1993;**5**:1-33 (in Turkish)

[6] Şengör AMC, Yılmaz Y. Tethyan evolution of Turkey: A plate tectonic approach. Tectonophysics. 1981;**75**:181-241

[7] Göncüoğlu MC, Dirik K, Kozlu H. Pre-alpine and alpine Terranes in Turkey: Explanatory notes to the Terrane map of Turkey. Ed.D. Papanikolaou, F.P. Sassi, IGCP project No:276 final volume: Terrane maps and Terrane descriptions. Annales Géologiques Des Pays Helléniques. 1997;**37**:515-536

[8] Başıbüyük Z, Yalçın H. Minerology, petrography and origin of hydrothermal alteration in eocene magmatites in central anatolia (Sivas-Turkey). Bulletin of The Mineral Research and Exploration. 2019;**158**:143-166

[9] Bingöl E. 1/2.000.000 ölçekli Türkiye Jeoloji Haritası. Maden Tetkik ve Arama Yayını, Ankara; 1989 (in Turkish)

[10] Tüysüz O, Karadeniz'den Orta Anadolu'ya Bir Jeotravers: Kuzey Neo-Tetisin Tektonik Evrimi. Türkiye Petrol Jeolojları Derneği Bülteni. 1993;**5**:1-33. (in Turkish)

[11] Görür N, Tüysüz O, Şengör AMC. Tectonic evolution of the Central Anatolian Basins. International Geology Review. 1998;**40**:831-850

[12] Seymen İ. Kaman (Kırşehir) dolayında Kırşehir masifinin stratigrafisi ve metamorfizması. TJK Bülteni. 1981;**24**(2):101-108 (in Turkish)

[13] Erler A, Bayhan H. Orta anadolu granitoidlerinin genel değerlendirilmesi ve sorunlari. Hacettepe Üniversitesi Yerbilimleri. 1995;**17**:49-67

[14] Otlu N. Kortundağ-Baranadağ Arası (D Kaman, KIRŞEHİR) Plütonik Kayaçlarının Petrolojik İncelenmesi; Cumhuriyet Üniversitesi Fen Bilimleri Enstitüsü; 1998 (in Turkish)

[15] Boztuğ D, Güney Ö, Heizler M, Jonckheer RJ, Tichomirowa M, Otlu N. 207Pb-206Pb,40Ar-39Ar and fission-track geothermochronology quantifying cooling and exhumation history of the Kaman-Kırşehir region intrusions, Central Anatolia, Turkey. Turkish Journal of Earth Sciences. 2009;**18**(1):85-108

[16] Kara H, Dönmez M. 1:100.000 ölçekli açınsama nitelikli Türkiye jeoloji Haritaları serisi, Kırşehir–G 17 paftası. MTA yayınları; 1990 (in Turkish)

[17] Otlu N, Boztuğ D. The coexistence of the silica oversaturated (ALKOS) and undersaturated alkaline (ALKUS) rocks in the Kortundağ and Baranadağ plutons from the central Anatolian alkaline plutonism, E Kaman/NW

Kırşehir, Turkey. Turkish Journal of Earth Sciences. 1998;7:241-257

[18] Ilbeyli N. Field, petrographic and geochemical characteristics of the Hamit alkaline intrusion in the central Anatolian crystalline complex, Turkey. Turkish Journal of Earth Sciences. 2004;**13**(3):269-286

[19] Deer WA, Howie RA, ve Russman J. Rock Forming Minerals. London: Longmans; 1965

[20] Lünel AT, Akman O. Hamitköy, Kaman, Kırşehir bölgesinde bulunan psodolosit oluşuğu ve basınç göstergesi olarak kullanımı. Maden Tetkik ve Arama Dergisi. 1985;**103**:103-104

[21] Streckeisen A. IUGS subcommission on the systematics of igneous rocks. Classification and nomenclature of volcanic rocks, lamprophyres, carbonatites, and melilite rocks. Recommendations and suggestions. Neues Jahrbuch für Mineralogie Abhandlungen. 1978;**134**:1-14

[22] Bayhan H. Geochemistry and genetic interpretation of alkaline rocks in the Bayındır-Akpınar (Kaman) area. Geological Bulletin of Turkey. 1988;**31**:59-70

[23] Le Maitre RW, Bateman P, Dudek A, Keller J, Lameyre J, Le Bas MJ, et al. Classification of Igneous Rocks and Glossary of Terms: Recommendations of the International Union of Geological Sciences Subcommission on the Systematics of Igneous Rocks. Oxford: Blackwell Scientific; 1989

[24] Kıymet D, Kadıoğlu YK. Nefelin siyenitlerin seramik sanayinde kullanılma potansiyeli: Buzlukdağ örneği. Pamukkale Üniversitesi Mühendislik Bilimleri Dergisi. 2018;**24**(6):1209-1219

Chapter 5

Mineralogical-Petrographical Investigation and Usability as the Gemstone of the North Anatolian Kammererite, Tokat, Turkey

İlkay Kaydu Akbudak, Zeynel Başibüyük and Gökhan Ekincioğlu

Abstract

Kammererite formations were observed in the region of Tokat province in the north of Anatolia. Kemmererite (purple, reddish, pink color) is present in the form of nodules or veins in chromium levels found in Mesozoic basic-ultrabasic rocks. In the surveys, it was found that archerite minerals do not show a widespread distribution and have different shades of pink and color and glassy brightness. Thin-section analyses were performed from kammererite samples. In the investigations, kammererite mineral showed brownish or pinkish pleochroism in plane-polarized light. In crossed polars, it was observed that they had interference color in grayish tones. Due to its low hardness, kammererite was treated with epoxy to increase its durability. In addition, it has been determined that they can be used in both jewelry and ornamental objects with the applied cabochon cutting styles.

Keywords: kammererite, gemstone, mineralogy, gemology, North Anatolia

1. Introduction

Kammererite mineral, which is rarely found in the world, is found in chromite deposits as reddish pink or purple-violet [1] colors as transparent or semitransparent. Its hardness is around 2.5, and its specific weight is 2.645 g/cm^3 [2, 3].

Kammererite mineral is one of the clinochlorine members of chlorite group in phyllosilicates. Clinochlore, which is one of the most common members of the chlorite group minerals [4], can be divided into three subvarieties according to body colors and implicational abundance of the main cations [5–11]. These are blackish-green or bluish-green colored clinochlore (ferroan clinochlore) [12], yellowish-green or green colored clinochlore (magnesian clinochlore) [13, 14], and magenta colored clinochlore (chromian clinochlore) [5, 11, 15–17]. In fact, it is well-known that the name clinochlore derived from "clino," which refers to the inclined optical axes and the Greek "chloros," for "green," its most typical color [4, 11, 18, 19].

Chromian clinochlore (kammererite) represented by the formula [$Mg_5(Al,Cr,Fe)_2 Si_3O_{10}(OH)_8$] [8, 9] is a hydrous silicate with a monoclinic IIb-2 polytype, with symmetry C^2 = m, and is extremely rare and of high interest for mineral collectors [20].

IntechOpen

Worldwide occurrences of chromian clinochlore (kammererite) in addition to Turkey are as follows: Australia (Coobina chromite mine, Sylvania Station, Meekatharra Shire, Western Australia), Austria (Gulsen, Sommergraben, Lobminggraben, Leoben, Styria), Ethiopia (Tumut River, Sosua Region, Benishangul-Gumaz Province), Finland (Elijarvi Cr Mine, Kemi, Lapland Region), Greece (Nea Roda, Chalkidiki Prefecture Macedonia), Italy (Locana, Orco Valley, Torino Province, Piedmont), Japan (Akaishi Mine, Ehime Prefecture, Shikoku Island), Russia (Poldnevaya village, Sverdlovsk Oblast, Middle Urals), and the United States, (Dunsmuir, Siskiyou Co., California; Cecil Co., Maryland; Green Mountain Mine, Day Book, Yancey Co., NC; Jackson Co., Oregion; Woods Chrome Mine, Texas, Little Britain Township, Lancaster Co., PA) [20].

2. Material and methods

Samples were taken from the study area in order to determine the distribution, the paragenetic relationships, and the mineralogical, geochemical, and gemological characteristics of kammererite. It has been engraved on 1/25000 map. Thin sections were prepared from kammererite and side rock samples taken from the field in thin-section laboratory of Kırşehir Ahi Evran University Geological Engineering Department. Mineralogical determinations (mineral paragenesis) were carried out by examining these samples under a polarizing microscope in Kırşehir Ahi Evran University Geological Engineering Mineralogy-Petrography Laboratory.

In addition, gem-cutting techniques were applied to the kammererite samples taken from the field by using diamond coating saw, sinter diamond abrasive discs and polishing machine, and the usability of kammererites as a gemstone was present.

Kammererite samples taken from the study area applied gem-cutting techniques in the Gemology Laboratory of Mersin University, School of Jewelry Technology and Design.

First, slices of coarse material were taken on the large cutting machine, and different shapes were marked. Edge trimming was done on the small cutting machine, and curves were made on the cabochon machine. Finally, abrasive and polishing processes were carried out to form cabochon stones. Because kammererite are fine grains and fine veins, it cannot be processed alone. For this reason, it was worked together with the side rock. The obtained gems can be used in jewelry as necklaces, rings, earrings, bracelets, brooches, and functional goods such as keychains.

Treatment studies of kammererite samples were carried out in the natural stone analysis laboratory of Kaman Vocational School of Kırşehir Ahi Evran University. First, the samples were kept in the oven at 75°C for one day to allow them to completely exhale. Then, the hot samples were kept in the mixture of epoxy and hardener for 1 day. As a result, the epoxy penetrated the capillary cavities of the samples, and the samples had a solid structure. Epoxy-treated specimens were processed using cabochon and simple step cutting methods.

3. Geology

The study area is located in the Middle Pontid Tectonic Belt [21], south of Tokat province.

Tokat metamorphites [22], which represent the oldest unit in the study area and contain schist, phyllite, marble, and metabasites, are Upper Paleozoic-Triassic and

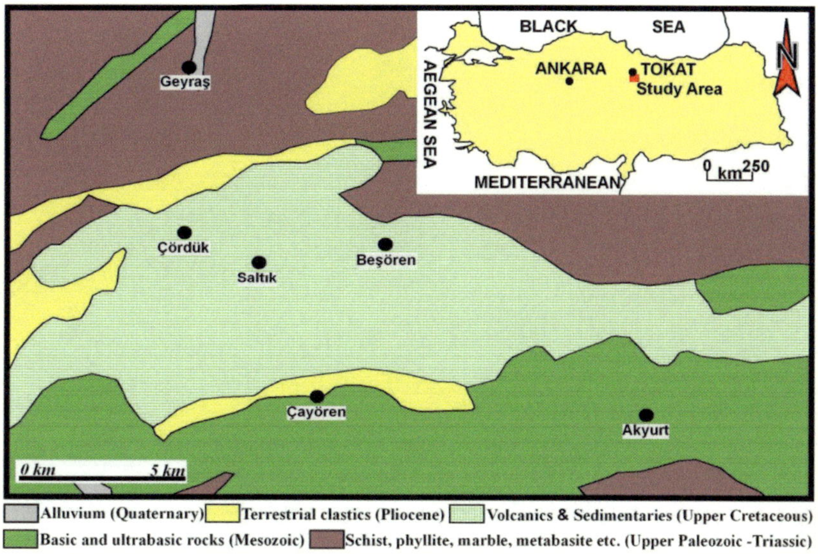

Figure 1.
Geological map of the study area [23].

are overlain by Mesozoic basic and ultrabasic rocks which are part of the ophiolitic series (**Figure 1**). These basic and ultrabasic rocks are overlain by Upper Cretaceous volcanic and sedimentary units. The youngest units in the study area are Quaternary alluviums.

4. Findings

4.1 Field studies

Kammererite formations in the region including Beşören and Saltık Villages within the borders of Tokat province in northern Anatolia are in purple-violet and reddish pink color and are in the form of nodules (**Figure 2a, b, d, e**) or veins (**Figure 2c, f**) within the chromium levels within the Mesozoic basic-ultrabasic rocks reaching up to 40 cm (**Figure 3**).

4.2 Mineralogical-petrographical-gemological investigations

Thin sections prepared in order to determine the mineral associations and textural relationships of the rock samples taken from the study area were examined under a polarizing microscope. Chromium minerals are observed as black color in plane-polarized light and crossed polars because they are opaque minerals (**Figure 4**).

Kammererite minerals has microcrystalline size. While the colorless, grayish, brownish, and pinkish pleochroism was observed in the plane-polarized light in the kammererite mineral (**Figure 4b, d, f, h**), the interference colors in black and white gray tones were observed in the crossed polars (**Figure 4a, c, e, g**).

It was observed in the surface investigations that kammererite minerals did not show a widespread distribution. Kammererite minerals in the study area have different shades of pink color and glassy brightness and are either transparent or semitransparent.

Figure 2.
View of kammererites in the field (kammererite nodules—a, b, d, and e; kammererite veins—c and f).

Figure 3.
Close-up view of kammererite samples taken from the study area.

First, kammererites were processed without any treatment (**Figure 5a**). They have low durability and very fine grain mineral composition. For this reason, treatment has been made in kammererite. Gemmologically better products were obtained (**Figure 5b**).

5. Results

Kammererites in the study area are in the form of nodules or thin veins at chromium levels within the basic-ultrabasic rocks of the Mesozoic age. As a result of

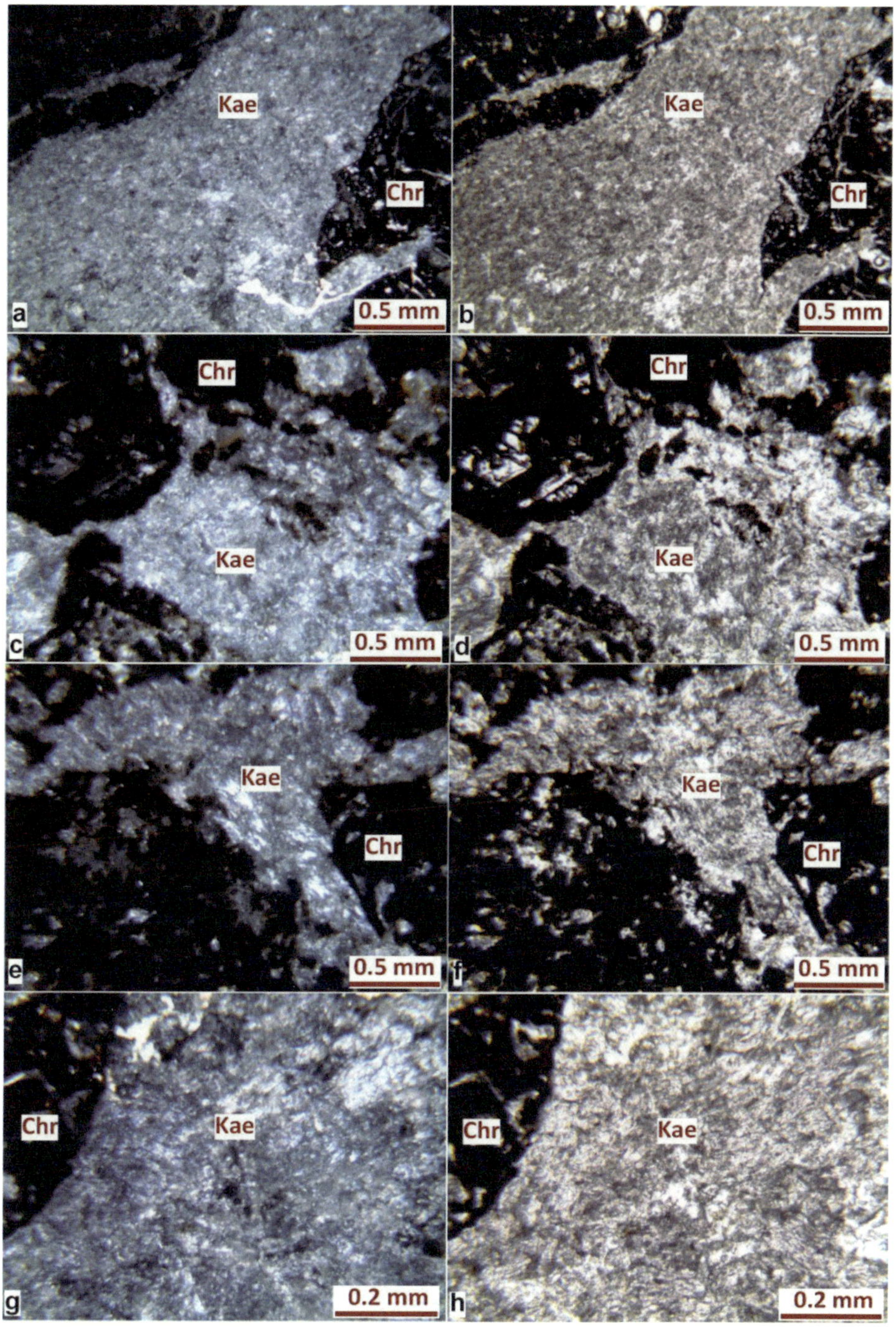

Figure 4.
Combination of chromium (Chr) and kammererite (Kae) minerals (a, c, e, g—crossed polars; b, d, f, h—plane-polarized light).

the surface study carried out in the study area, it is observed that the kammererites formation does not show much spread. They are found in different shades of pink in the field with abundant cracked cracks. Thin sections made from kammererite samples taken from the field are colorless, grayish brownish, and pinkish pleochroism in plane-polarized light. In crossed polars, interference colors are observed in

Figure 5.
Gemstones made from kammererite samples taken from the study area (a—processed samples without treatment; b—processed samples after treated with epoxy).

black and white gray tones. The opaque minerals that are impermeable to light are the chromium minerals.

Bir mineralin süstaşı olarak kullanılabilmesi için temel özelliklerden olan, nadir bulunma, dayanıklılık, güzellik (renk, saydamlık vb.), işlenebilirlik özelliklerini barındırması beklenmektedir. Inceleme alanındaki Kemereritler nadir bulunma, güzellik ve işlenebilirlik özelliklerine sahiptir. Düşük dayanıklılığı ise iyileştirme yöntemleri ile arttırılabilir. Rarity, durability, beauty (color, transparency, etc.), and processability which are the basic properties of gemstones are expected from a mineral for being used as gemstones. Kammererites in the study area have rarity, beauty, and processability properties. Its low durability can be increased by treatment methods.

As a result of the lapidary studies made from the samples taken from the study area, it was observed that kammererite minerals can be processed together with the side rock and used in jewelry and ornamental production. However, the low hardness of kammererites, while facilitating workability, adversely affects their durability. For this reason, after the treatment (with epoxidation method), both increased durability and visually appealed.

Considering the rarity of studies related to the rarity of kammererite in the world, this study is also important in terms of providing resources for those working and researching in this field.

Acknowledgements

This study was carried out within the scope of the project of MMF.A4.18.014 supported by Kırşehir Ahi Evran University Scientific Research Projects Coordination Unit. We would like to extend our thanks to the Kırşehir Ahi Evran University Scientific Research Projects Coordination Unit, which provided financial support to carry out this work.

Conflict of interest

The authors declare no conflict of interest.

Author details

İlkay Kaydu Akbudak[1], Zeynel Başibüyük[1] and Gökhan Ekincioğlu[2]*

1 Faculty of Engineering and Architecture, Department of Geological Engineering, Kirşehir Ahi Evran University, Kirşehir, Turkey

2 Department of Mining and Mineral Extraction, Kaman Vocational School, Kirşehir Ahi Evran University, Kirşehir, Turkey

*Address all correspondence to: gokhanekincioglu@gmail.com

References

[1] Temur Y. Süs Taşlari. SDUGEO e-dergi. 2011;**1**:9-21. ISSN: 1309-6656

[2] Aydın ŞN, Bektur Z, ve Çelebioğlu FN. MTA Tabiat Tarihi Müzesinde Sergilenen Mineraller. Ankara: Yayınlarından/Maden tetkik ve arama enstitüsü; 1998. (in Turkish)

[3] Şahin M, Koşun E, Ağrılı H, Mengi H. Mineraller. Ankara: Yayınlarından/Maden tetkik ve arama enstitüsü; 2000

[4] Back ME, Mandarino JA, Fleischer M. Fleischer's Glossary of Mineral Species. 10th ed. Tucson: Mineralogical Record Inc.; 2008

[5] Brown BE, Bailey SW. Chlorite polytypism: II. Crystal structure of a one-layer Cr-chlorite. American Mineralogist. 1963;**48**:42-45

[6] Bailey SW. Chlorites: Structures and crystal chemistry. In: Bailey SW, editor. Hydrous Phyllosilicates (Exclusive of Micas), Reviews in Mineralogy and Geochemistry. Vol. 19. Chantilly, Virginia: Mineralogical Society of America; 1988. p. 347.403

[7] Grevel K, Fasshauer DW, Erzner S. New compressibility data for clinochlore, kyanite, Mg-chloritoid, and Mg-staurolite. European Journal of Mineralogy. 1997;**1**:138

[8] Joswig W, Fuess H, Mason SA. Neutron diffraction study of a one-layer monoclinic chlorite. Clays and Clay Minerals. 1989;**37**(6):511-514

[9] Zheng H, Bailey SW. Structures of intergrown tricilinic and monoclinic IIb chlorites from Kenya. Clays and Clay Minerals. 1989;**37**(308):316

[10] Theye T, Parra T, Lathe C. Room temperature compressibility of clinochlore and chamosite. European Journal of Mineralogy. 2003;**15**(3):465-468

[11] Hatipoğlu M, Oğuzer MB, Buzlu HB. Gemmological and mineralogical investigations and genesis of the Kammererite from the Keşiş (Erzincan) and Kop (Erzurum) mountains. Journal of African Earth Sciences. 2013;**84**:20-35

[12] Rule AC, Bailey SW. Refinement of the crystal structure of a monoclinic ferroan clinochlore. Clays and Clay Minerals. 1987;**35**:129-138

[13] Hayes JB. Polytypism of chlorite in sedimentary rocks. Clays and Clay Minerals. 1970;**19**:285-306

[14] Welch MD, Kleppe AK, Jephcoat AP. Novel high-pressure behavior in chlorite: A synchrotron XRD study of clinochlore to 27 GPa. American Mineralogist. 2004;**89**:1337-1340

[15] Dietrich R, Medenbach O. Kämmererite from the Kop krom mine, Kop Dağları (Turkey). The Mineralogical Record. 1978;**9**:277-287

[16] Chadwick KM. Chromium-rich clinochlore (Kammererite) from Turkey. Gems & Gemology. 2008;**44**:168-169

[17] Farges F. Chromium speciation in oxide-type compounds: Application to minerals, gems, aqueous solutions and silicate glasses. Physics and Chemistry of Minerals. 2009;**36**:463-481

[18] Schumann W. Gemstones of the World. Sterling Publishing Company Inc., Inc. 2009

[19] Arem JE. Color Encyclopedia of Gemstones. Van Nostrand Reinhold Company; 1987

[20] Hatipoğlu M. Photoluminescence response from the chromian clinochlore (Kammererite). Spectroscopy Letters. 2014;**47**(10):746-753

[21] Yolcubal HG, Akyazı M, Sezen TF, Toprak Ö, Yasin M, Canpolat FK, et al. Turhal-Pazar-Zile (Tokat) Yöresinin Üst Mesozoyik Stratigrafisi. Geological Bulletin of Turkey. 2014;**57**(1)

[22] Yılmaz A, And Yılmaz H. Geology and structural evolution of the Tokat Massif (Eastern Pontides, Turkey). Turkish Journal of Earth Sciences. 2004;**13**:231-246. Copyright ©TÜBİTAK

[23] MTA. 1/500.000 Türkiye Jeoloji Haritası. Ankara: Maden Tetkik ve Arama Genel Müdürlüğü; 2002

Chapter 6

Enhanced Humidity Sensing Response in Eu^{3+}-Doped Iron-Rich $CuFe_2O_4$: A Detailed Study of Structural, Microstructural, Sensing, and Dielectric Properties

I.C. Sathisha, K. Manjunatha, V. Jagadeesha Angadi, B. Chethan, Y.T. Ravikiran, Vinayaka K. Pattar, S.O. Manjunatha and Shidaling Matteppanavar

Abstract

The $CuFe_{(2-x)}Eu_xO_4$ (where x = 0.00, 0.01, 0.02, 0.03) nanoparticles are synthesized by solution combustion method. The influence of Eu^{3+} on the structural, morphological, dielectrical, and humidity sensing study is recorded. The XRD pattern peaks of the as-prepared $CuFe_{(2-x)}Eu_xO_4$ (where x = 0.00, 0.01, 0.02, 0.03) nanoparticle confirm the polycrystalline spinel cubic structure with a small amount of CuO impurity phase at 38.87° and 48.96°. Surface morphology of the samples was studied by scanning electron microscope (SEM) images of the nanoparticles, and their respective average grain size was estimated using Image software. Chemical composition of all prepared samples was analyzed by EDS spectra. The dielectric parameters of AC conductivity, electric modulus, and impedance of the samples were measured over a range of frequencies from 0.1 KHz to 1 MHz at room temperature. Europium-doped copper ferrite samples showed good humidity sensing response, response and recover times, and stability over a %RH range of 11–91%. These types of samples are very useful for sensor application, battery applications, electronic applications, and automotive applications.

Keywords: ferrite, Eu-doping, dielectric, impedance, humidity sensing

1. Introduction

In recent years, the great developments in the applications of larger ionic rare earth-doped ferrites have been effectively realized in many prominent fields mainly in sensors, communication, and electronics. Copper ferrite metal oxide is a *p*-type semiconductor metal oxide and has been used in organic synthesis as a catalyst. Recently, many research groups have explored the importance of copper ferrite metal oxide in wide advanced applications because of their unique chemical, biocompatible, and physical properties [1]. The $CuFe_2O_4$ ferrite has spinel cubic structure with general formula Fd3m. The $CuFe_2O_4$ ferrite belongs to a general formula AB_2O_4 in which A is a divalent ion site occupied by the Cu^{2+} ions and B is a trivalent ion site occupied by Fe^{3+} ions. The rare earth-doped $CuFe_2O_4$ ferrites have been used in many fields such as electronic devices, drug delivery systems, cancer therapy, and magnetic recording [1]. Rare earth-doped ferrites are also useful in high voltage electronics due to its negligible eddy current losses, high electrical resistivity, high permeability, magneto-optical, and magnetoresistive properties [2–4]. Generally, the trivalent ions are lesser in size than a divalent ion, and hence exchange of cations among the A and B sites plays a vital role in studying the structural, morphological, dielectric, and humidity sensing behavior of spinel copper ferrites. Copper ferrite has considerable good attraction of potential applications in various devices, like cores in transformers and microwave absorption [3, 4]. Copper ferrite nanoparticles can be prepared by various techniques like sol-gel method, coprecipitation method, solution combustion method, etc. [5, 6]. Many researchers have explored the properties of spinel ferrites by doping with different larger rare earth ions like samarium, terbium, gadolinium, and cerium in it. The europium-doped copper ferrites can be used in high-frequency applications due to the rare earth-doped ferrites showing low dielectric loss, good sensing response, and low value of conductivity. Europium-doped copper ferrites are usually considered as inverse spinel ferrites.

In the present work, spinel copper ferrites were prepared by solution combustion method and then in Eu^{3+} ions will be incorporated to investigate its structural, morphological, dielectric, and humidity sensing behavior at room temperature.

2. Experimental details

2.1 Required oxidizers and fuels

The required oxidizers (metal nitrates), viz., copper nitrate $[Cu(NO_3)_2{\cdot}3H_2O]$, europium nitrate $[Eu(NO_3)_3{\cdot}5H_2O]$, and ferric nitrate $[Fe(NO_3)_2{\cdot}9H_2O]$, and reducing agents, viz., carbamide $[NH_2CONH_2]$ and glucose $[C_6H_{12}O_6]$, all were purchased from S.D. Fine Chemicals, Mumbai, India.

2.2 Method

The $CuFe_{(2-x)}Eu_xO_4$ (where x = 0.00, 0.01, 0.02, 0.03) nanoparticles were synthesized by solution combustion method by using stoichiometry amount of copper nitrate, europium nitrate, and ferric nitrate as metal nitrates and carbamide and glucose as fuels. The weighed metal nitrates and fuels were taken in 250 ml borosil glass beaker; then all nitrates and fuels were diluted with distilled water and kept on magnetic stirrer about 45 min to obtain a homogeneous solution. This solution was kept in preheated muffle furnace at $450 \pm 10°C$ temperature to ignite, to get a copper ferrite nanoparticle. Obtained $CuFe_{(2-x)}Eu_xO_4$ (where x = 0.00, 0.01, 0.02, 0.03) nanoparticles were taken in to mortar and then pestle it for getting fine powder of nanoparticles. The solution combustion technique flowchart for europium-doped copper ferrite as shown in **Figure 1**.

2.3 Characterizations

For as-synthesized $CuFe_{(2-x)}Eu_xO_4$ (where x = 0.00, 0.01, 0.02, 0.03) nanoparticles, X-ray diffraction characterization was carried out by using

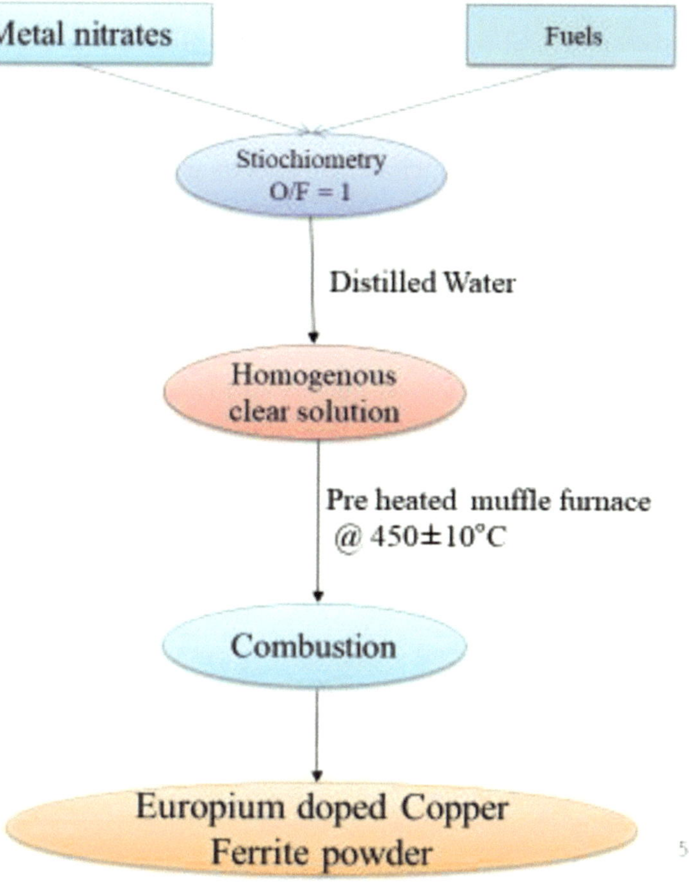

Figure 1.
Flowchart to show solution combustion technique of synthesizing europium-doped copper ferrite.

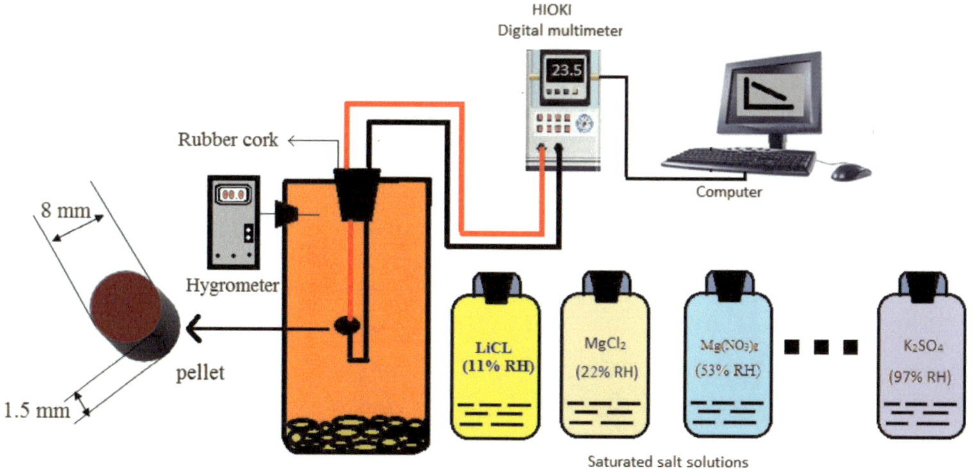

Figure 2.
Schematic illustration of humidity sensing setup.

Cu-K_α radiation of wavelength 1.5404 Å and the diffractogram run with two theta (2θ) range from 10° to 80°. The XRD data were refined by using full proof software. The Rietveld refined XRD pattern peaks confirm the polycrystalline spinel cubic structure. Scanning electron microscopy (SEM) and energy-dispersive X-ray spectroscopy (EDS) were carried out by using JEOL (Model JSM-840) instrument respectively to understand structural morphology and elemental analysis. The mean grain size of the particles was estimated from SEM micrographs by using ImageJ software. From SEM and EDS to understand structural morphology and elemental analysis. The AC eclectic parameters were measured over a range of frequencies from 50 Hz to 10 MHz using Wayne Kerr 6500B series impedance analyzer. For humidity sensing AC conduction studies, powder samples were pressed in the form of pellet under the hydraulic pressure of 5 tons. Further, two faces of these pellets were painted with silver paste for electrical contact. The pellet was held between two probes and then placed in a glass chamber through a rubber cork and the other end of the electrodes connected to the programmable HIOKI Digital Multimeter to record the resistance corresponding to various RH relative humidities from 11% RH to 97%. The schematic illustration of humidity sensing setup is shown in **Figure 2**.

3. Results and discussion

3.1 XRD analysis

Figure 3 shows the Rietveld refined XRD patterns of as-synthesized $CuFe_{(2-x)}Eu_xO_4$ (where x = 0.00, 0.01, 0.02, 0.03) nanoparticles. XRD patterns of all samples show the polycrystalline spinel cubic structure with space group Fd3m with a small amount of CuO impurity phase. The indexed XRD peaks are well

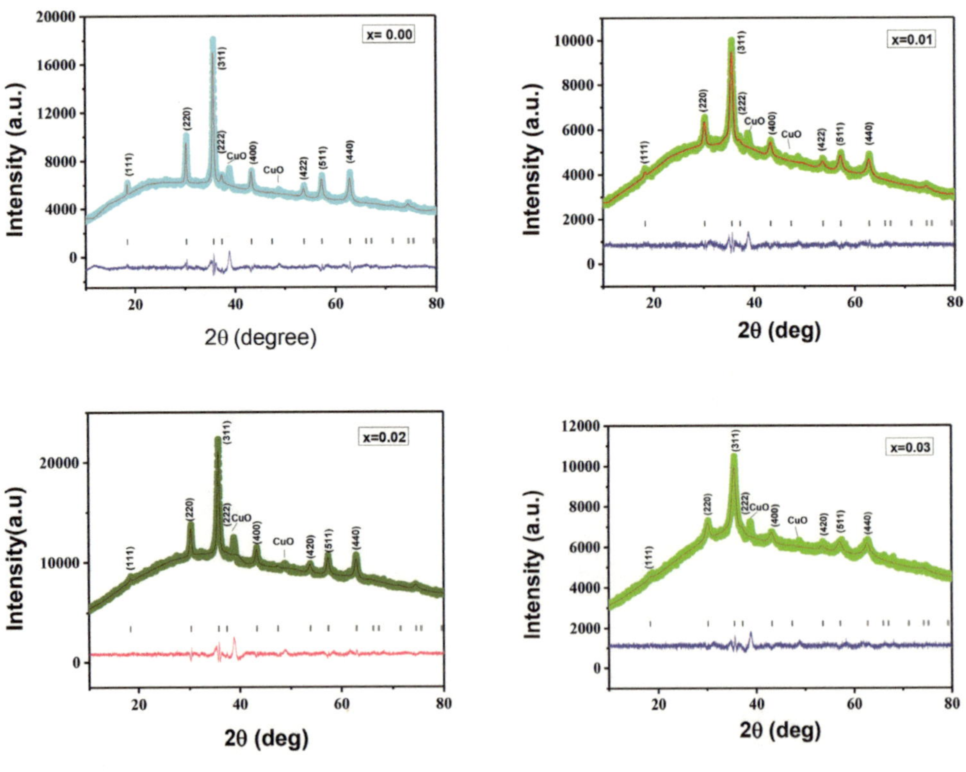

Figure 3.
Rietveld refined XRD patterns of the $CuFe_{(2-x)}Eu_xO_4$ (where x = 0.00, 0.01, 0.02, 0.03) nanoparticles.

matched with spinel cubic structure of JCPDS card number 74-2400 and the CuO impurity peaks observed at 38.87° and 48.96° (JCPDS 80-1268) [7, 8]. The lattice parameter is found to increase with increase in Eu^{3+} concentration due to difference in ionic radius. The ionic radii of Fe^{3+} (0.67 Å) are lesser than that of Eu^{3+} (0.99 Å) ions; this confirms the occupancy of europium on an octahedral site. The average crystallite size of $CuFe_{2-x}Eu_xO_4$ (where x = 0.00, 0.01, 0.02, 0.03) nanoparticles estimated from the Debye-Scherrer relation and the average particle size of all the samples were found to be in the range of 16–51 nm [9]. The strain values are calculated by using equation $\varepsilon = {}^{\beta cos\theta}/_{4}$ and are tabulated in **Table 1**. Further, the hopping length of L_A and L_B is calculated, the hopping length increasing with increase in Eu^{3+} concentration. This happened because Fe^{3+} ions are replaced by the relative number of Eu^{3+} in octahedral site and the variation of hopping length of tetrahedral site (L_A) and octahedral site (L_B) is shown in **Table 1**.

3.2 SEM and EDS

The surface morphology and elemental analysis of the $CuFe_{(2-x)}Eu_xO_4$ (where x = 0.00, 0.01, 0.02, 0.03) nanoparticles were performed with scanning electron microscope and energy-dispersive X-ray spectroscope. From **Figure 4(left)** we can clearly see that all the particles are spherical in shape and exhibit smooth surface with an average grain size of 20–40 nm. All samples of the SEM micrographs show highly porous nature, and the appearance of the dry foamy powder is due to the evolution of the gases during the combustion process [10]. **Figure 4(right)** shows the EDS spectrum of $CuFe_{(2-x)}Eu_xO_4$ (where x = 0.00, 0.01, 0.02, 0.03) nanoparticles. The $CuFe_{(2-x)}Eu_xO_4$ (where x = 0.01, 0.02, 0.03) of EDS spectrum depicts Cu, Fe, Eu, and O peaks are clearly seen with Cuo impurity peaks (**Figure 4(right)**). $CuFe_{(2-x)}Eu_xO_4$ (where x = 0.00) of EDS spectrum depicts Cu, Fe, and O peaks are clearly seen with Cuo impurity peak. The Eu^{3+} peak appeared in all samples except when x = 0, and its intensity of Eu^{3+} peak increase with europium concentration increases. The grain size distribution histogram of $CuFe_{(2-x)}Eu_xO_4$ (where x = 0.00, 0.01, 0.02, 0.03) nanoparticles is shown in **Figure 4(left)**, and the average grain size of $CuFe_{(2-x)}Eu_xO_4$ (where x = 0.00, 0.01, 0.02, 0.03) nanoparticles is estimated from SEM micrographs using ImageJ software. The average grain sizes of all the particles are well matched with the crystallite size as shown in **Table 1**.

3.3 Dielectric studies

3.3.1 Variation of frequency with AC conductivity

Figure 5 shows the variation of frequency with AC conductivity plots, respectively, for $CuFe_{(2-x)}Eu_xO_4$ (where x = 0.00, 0.01, 0.02, 0.03) nanoparticles

Eu^{3+} content	Lattice parameters (Å)	Crystallite size D in nm	Volume (Å^3)	Strain Є (radian)	Space group	Average grain size	Hopping length (Å)	
							L_A	L_B
x = 0.0	8.126	25	539.29	1.38×10^{-3}	Fd3m	28	3.518	2.873
x = 0.01	8.131	16	540.23	2.14×10^{-3}	Fd3m	30	3.520	2.874
x = 0.02	8.143	21	542.71	1.63×10^{-3}	Fd3m	40	3.526	2.879
x = 0.03	8.154	51	544.91	6.98×10^{-3}	Fd3m	33	3.531	2.883

Table 1.
Structural parameters of the $CuFe_{(2-x)}Eu_xO_4$ (where x = 0.00, 0.01, 0.02, and 0.03) nanoparticles.

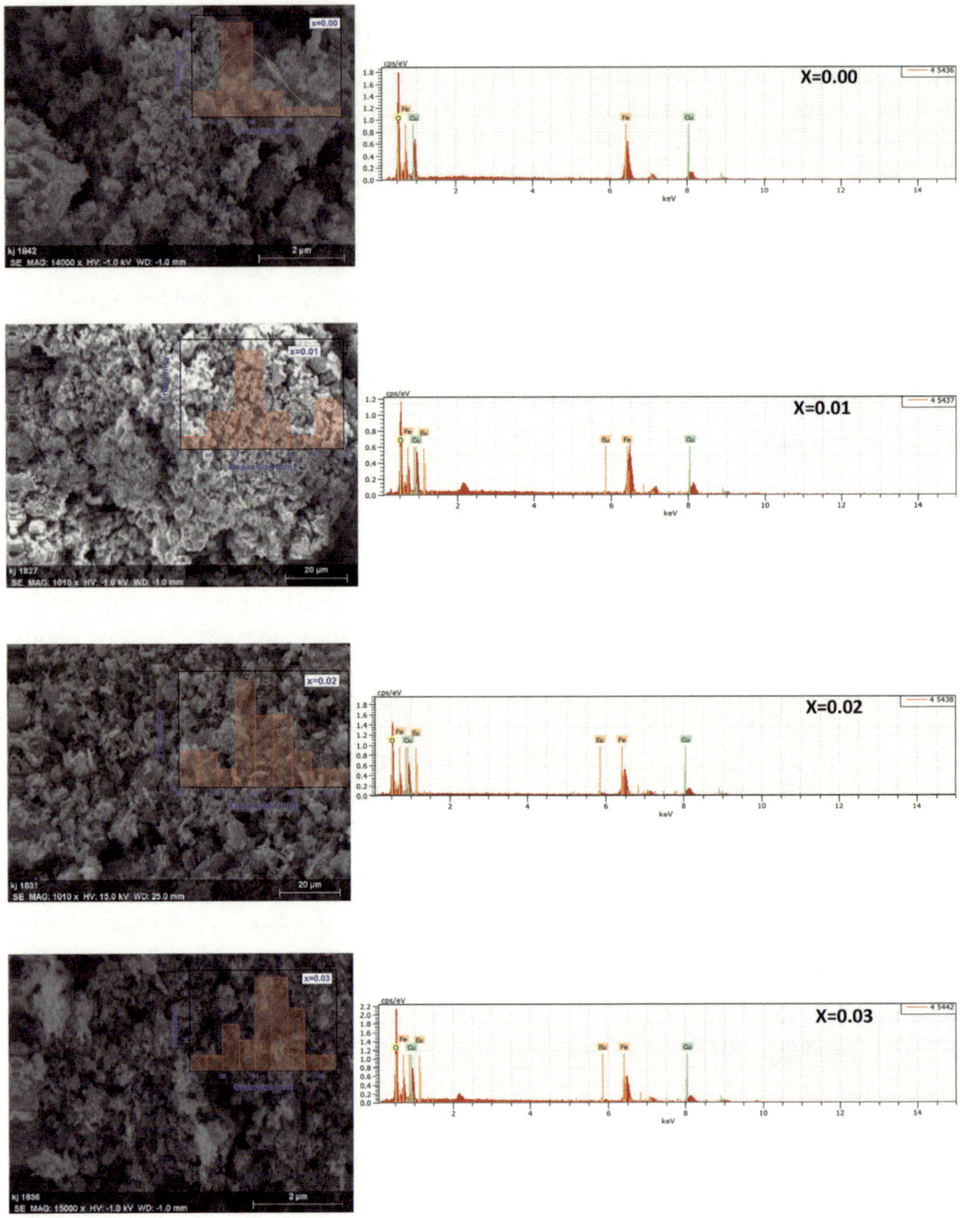

Figure 4.
(Left) SEM micrographs and (right) EDS spectra of the $CuFe_{(2-x)}Eu_xO_4$ *(where x = 0.00, 0.01, 0.02, and 0.03) nanoparticles.*

measured over the frequency range of 0.1 kHz to 1 MHz at room temperature. The figure clearly reveals the AC conductivity of each sample increases linearly with frequency. The exchange of electrons between A site and B site enhances the AC conductivity. The sample with a higher concentration of europium ions shows high values of AC conductivity. The substitution of europium ions on the copper ferrites obstructs the exchange of electrons between A sites and B sites, resulting in the decrease in the AC conductivity. At the lower-frequency region, the effect of grains and grain boundary is dominant, and it causes decreases in the exchange of electron between Fe^{2+} ions and Fe^{3+} ions, so small AC conductivity values have been observed. The polarization of all the samples of spinel copper ferrites is increased at high

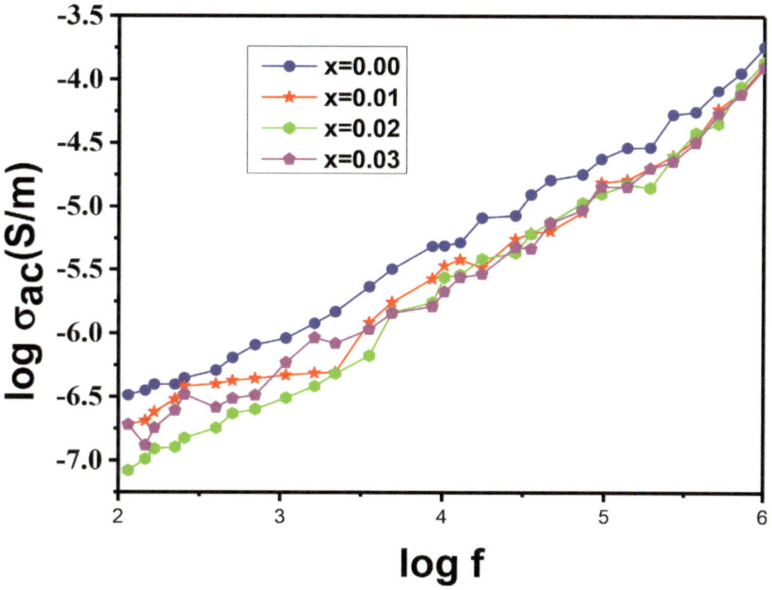

Figure 5.
The variation of frequency with AC conductivity for the $CuFe_{(2-x)}Eu_xO_4$ (where x = 0.00, 0.01, 0.02, and 0.03) nanoparticles.

frequency. In the copper ferrite, the part of the grains is very essential at greater frequency than grain boundary's part; it causes the conduction to be enhanced in spinel ferrites [11]. The doping of europium ions to the Fe^{3+} ions from the octahedral site disrupts the conduction mechanism, and the AC conductivity decreases. The maximum value of AC conductivity is found to be x = 0.00 concentration.

3.3.2 Variation of frequency with real and imaginary part of electric modulus

Figure 6 shows the variation of frequency with real and imaginary parts of electric modulus plots respectively for $CuFe_{(2-x)}Eu_xO_4$ (where x = 0.00, 0.01, 0.02, 0.03) nanoparticles. These were studied over the frequency range of 0.1 kHz to 1 MHz at room temperature. To study the character of grain boundary and grains over the frequency range of 0.1 kHz to 1 MHZ, electric modulus analysis was taken out. For this basis, the graphs plot between real and imaginary parts of electric

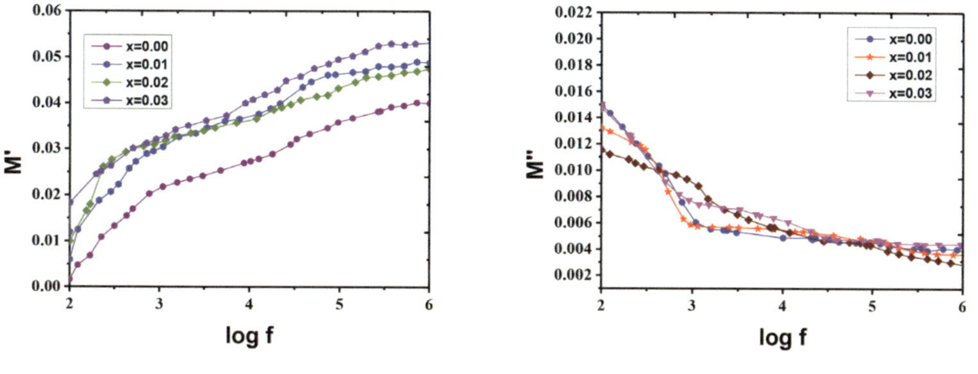

Figure 6.
The variation of frequency with real and imaginary parts of electric modulus for the $CuFe_{(2-x)}Eu_xO_4$ (where x = 0.00, 0.01, 0.02, and 0.03) nanoparticles.

modulus along the y-axis and the frequency along the x-axis were taken. In copper ferrites, the real parts of the electric modulus M' show very low M' values at lower-frequency region and increase linearly with the increase in the frequency of the external applied field [12]. In the case of the imaginary part of electric modulus (M''), high electric modulus at lower-frequency region is shown and again decreases with increase in the frequency of external applied field. The real and imaginary parts of electric modulus were maximum at x = 0.00 concentration.

3.3.3 Cole-Cole plots

Figure 7 shows the imaginary part of electric modulus with real part of electric modulus plots, respectively, for $CuFe_{(2-x)}Eu_xO_4$ (where x = 0.00, 0.01, 0.02, 0.03) nanoparticles at room temperature. To examine the effect of grain boundaries and grains within spinel ferrites, a plot can be drawn between real and imaginary parts of dielectric constant and dielectric loss or using the values of the real and imaginary parts of impedance, but no satisfactory results were obtained from the aforesaid formalisms. The figure plot between real and imaginary parts of dielectric modulus clearly shows the semicircles within the given range of these quantities. The Cole-Cole plots between the real part of the electric modulus and imaginary parts of the electric modulus can give good results. The role of grains and grain boundaries is very good because semicircles are observed at all the samples [9, 12]. The maximum peak of the diameter observed semicircle increases with the substitution of europium rare earth ions. The maximum intensity of peak is observed at x = 0.03 concentration.

Figure 7 shows Cole-Cole (the imaginary part of impedance with real part impedance) plots for $CuFe_{(2-x)}Eu_xO_4$ (where x = 0.00, 0.01, 0.02, 0.03) nanoparticles at room temperature.

In these impedance analysis plots, we observed only one semicircle clearly at x = 0.00 concentration, this study indicating the role of grain boundaries predominated, and the contribution from the grain was not resolved from this impedance analysis [13]. Sivakumar et al. reported a similar result from nanocrystalline cobalt ferrites [14]. The radii of the semicircles decreased with increasing concentration; this sign indicates a decrease in relaxation.

3.4 Humidity sensing studies

3.4.1 Variation of resistance and sensing response with relative humidity

In **Figure 8**, the curves are plotted between the relative humidity along the x-axis and resistance along the y-axis for $CuFe_{(2-x)}Eu_xO_4$ (where x = 0.00, 0.01, 0.02, 0.03) nanoparticles at room temperature. For all samples, the resistances are

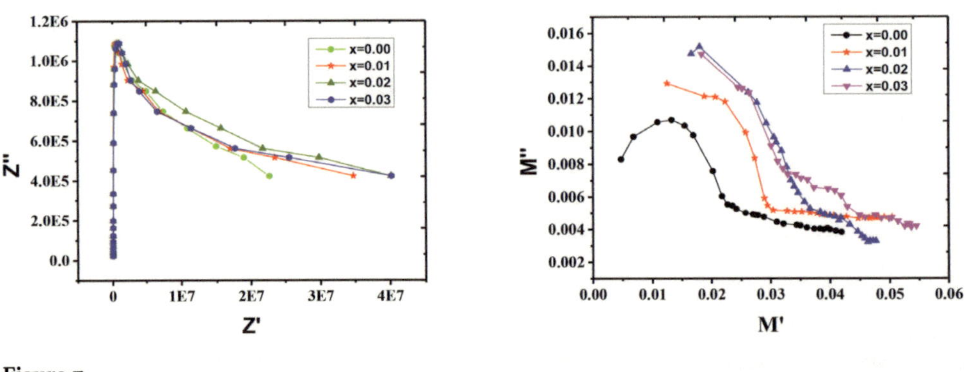

Figure 7.
Imaginary part of electric modulus with real part of electric modulus and the imaginary part of impedance with real part impedance, respectively, for $CuFe_{(2-x)}Eu_xO_4$ (where x = 0.00, 0.01, 0.02, and 0.03) nanoparticles.

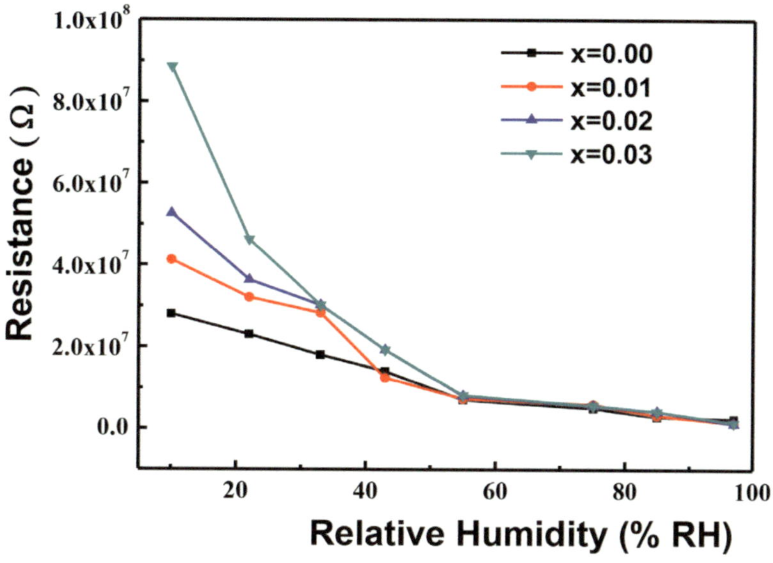

Figure 8.
The variation of resistance with relative humidity for $CuFe_{(2-x)}Eu_xO_4$ (where x = 0.00, 0.01, 0.02, and 0.03) nanoparticles.

decreased from 1×10^7 Ω to 100×10^7 Ω for relative humidity varying from 11 to 97% RH. From the figure, the maximum variation in resistance is observed for $CuFe_{(2-x)}Eu_xO_4$ (where x = 0.03) sample as compared to other samples. So, the humidity sensing response of each sample was calculated by using the following equation and plotted against %RH shown in **Figure 9**.

$$\text{Humidity sensing response} = \frac{\text{resistance of lower relative humidity} - \text{resistance at any relative humidity}}{\text{resistance of lower relative humidity.}} \quad (1)$$

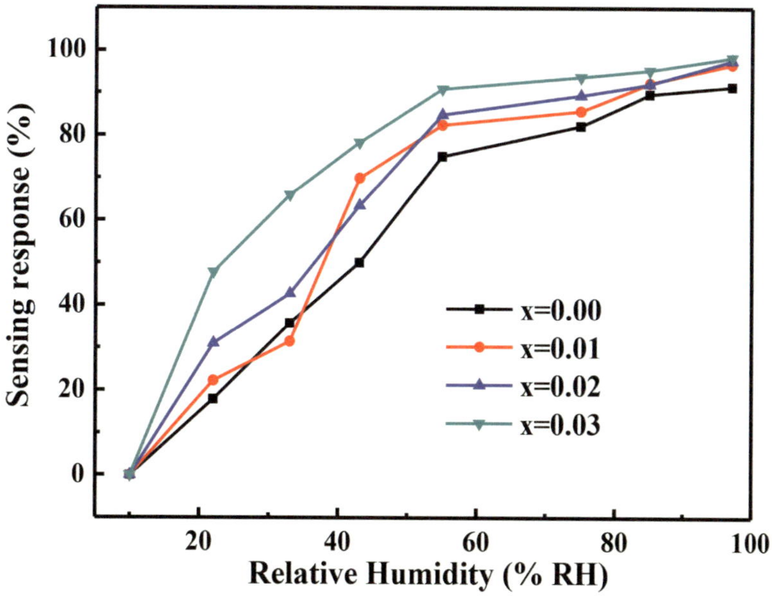

Figure 9.
The variation of sensing response (%) with relative humidity for $CuFe_{(2-x)}Eu_xO_4$ (where x = 0.00, 0.01, 0.02, and 0.03) nanoparticles.

The $CuFe_{(2-x)}Eu_xO_4$ (where x = 0.03) has shown the maximum sensing response; hence it is essential to know its sensing mechanism. The sensing mechanism is discussed on the basis of three sequential steps: chemisorptions, first step of physisorption, and second step of physisorption, followed by capillary condensation. In the initial stage, the water molecule adsorbed to the sensing surface and gets self-ionized to form H^+ and OH^- ions [15].

$$H_2O \Leftrightarrow H^+ + OH^- \quad (2)$$

In the beginning stage, the dissociated OH^- ions get attached to the sensing surface forming a chemisorption layer, and H^+ ions are released. These chemisorbed two OH^- ions form a hydrogen bond with the neighboring water molecule to form bulk water (H_3O^+). This forms the first physisorbed layer. Thus formed bulk water dissociates to form H_2O and H^+ ion. The released H^+ ions transfer from one water molecule to another through the braking and making of bonds. This is in accordance with the Grotthuss mechanism [16].

$$2H_2O \rightarrow H_3O^+ + OH^- \quad (3)$$

$$H_3O^+ \rightarrow H_2O + H^+ \quad (4)$$

As the RH increases, the physisorbed water molecules get piled up on one another forming the second step of physisorption. At the last stage, the adsorbed water molecule condenses in the capillary pores, leading to increase in the protonation. These results in the decrease in the resistance and in turn increase in its conductivity.

3.4.2 Sensing response and recovery

Nowadays, the sensing response and recovery time characteristics and stability testing are required for device fabrication of humidity sensing material. The response and recovery time were measured only for $CuFe_{(2-x)}Eu_xO_4$ (where x = 0.03) because of good sensing response as compared to other samples. In response and recovery time studies, we used two chambers, in that one chamber containing lower relative humidity of 11% RH another of higher relative humidity of 95% RH. The sensing response time of 63 s was recorded when the sample was moved from relative humidity of 11% RH to relative humidity of 97% RH, and the recovery time of 164 s was recorded when sample was moved from relative humidity of 97% RH to relative humidity of 11% RH (**Figure 10**). The difference between sensing response time and recovery time is small. These studies clearly show that the response and recovery time of Eu^{3+}-doped $CuFe_2O_4$ sample is slightly better than response and recovery time of nickel copper-zinc ferrite synthesized by coprecipitation method [17].

3.4.3 Stability

Stability testing is one of the important tests for practical application in device fabrication of sensing material. For stability testing, $CuFe_{(2-x)}Eu_xO_4$ (where x = 0.03) sample pellet was tested at relative humidity of 97% RH and 55% RH for every 10 days in 2 months. **Figure 11** shows the stability curves at 55% RH and 97% RH for $CuFe_{(2-x)}Eu_xO_4$ (where x = 0.03) sample at room temperature. The figure clearly indicates that both relative humidity samples show highly stable response at room temperature during that period. So, sample shows that humidity sensing material is highly stable at larger concentration of europium-doped copper ferrite at

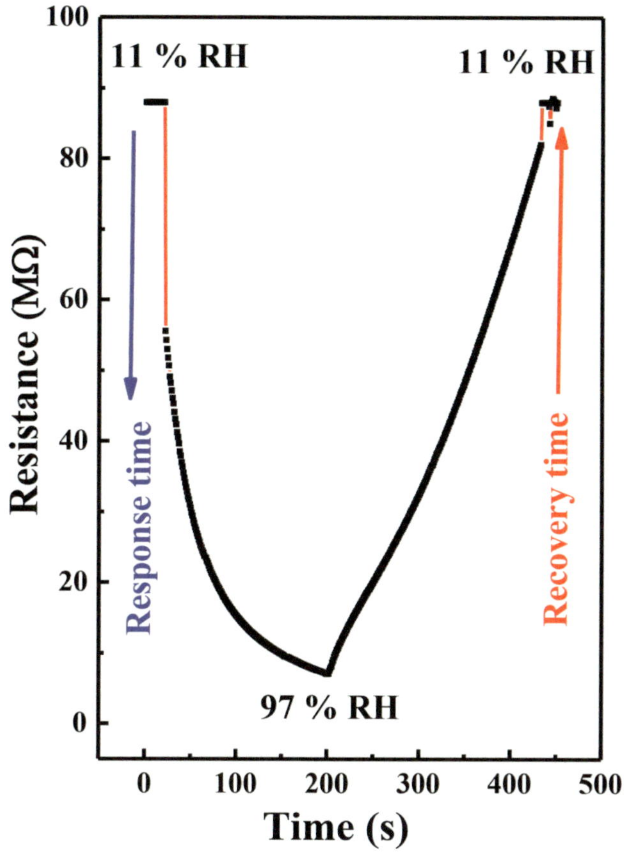

Figure 10.
The sensing response and recovery characteristic for $CuFe_{(2-x)}Eu_xO_4$ (where x = 0.00, 0.01, 0.02, and 0.03) nanoparticles.

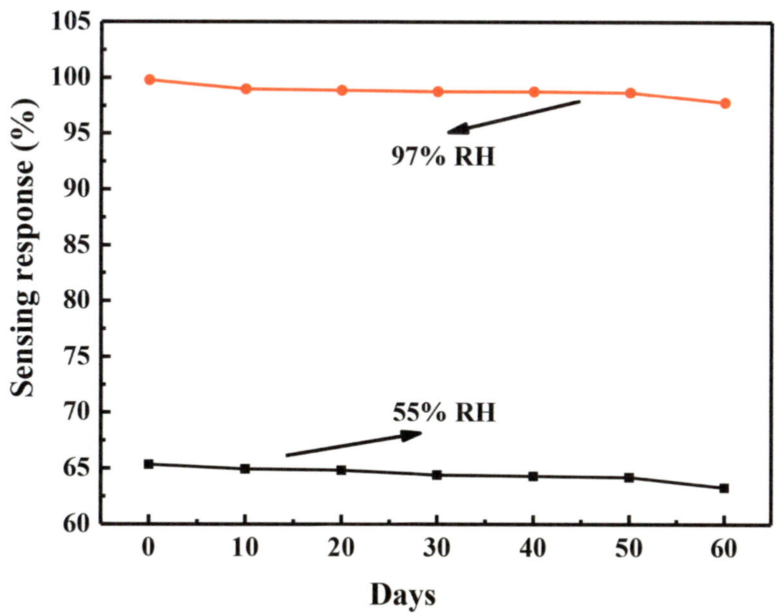

Figure 11.
The humidity sensing stability characteristic for $CuFe_{(2-x)}Eu_xO_4$ (where x = 0.00, 0.01, 0.02, and 0.03) nanoparticles.

room temperature and this is the proof for good practical application. The less concentration of europium-doped copper ferrite sample was not tested for stability because at lower concentration samples show less sensing response compared to higher concentration. However low sensing response ferrites also have good potential applications such as electronic and battery applications [17, 18].

4. Conclusions

The $CuFe_{(2-x)}Eu_xO_4$ (where x = 0.00, 0.01, 0.02, 0.03) nanoparticles were synthesized by solution combustion method using mixture of fuels for the first time. XRD patterns reveal the polycrystalline spinel cubic structure having space group Fd3m with small amount of impurity phases at 38.87° and 48.96°. The SEM micrograph reveals the formation of dry foamy powders during combustion process and porous nature of the sample. The average grain size of the samples was measured from SEM micrographs. The chemical composition was analyzed by using EDS analysis. The dielectric parameters vary with frequency. In the copper ferrite, the role of the grains is very essential at greater frequency as compared to grain boundary's role. The radii of the semicircles decreased with increasing concentration; this sign indicates a decrease in relaxation. The humidity sensing response at higher concentration is very good as compared to other concentration. The response and recovery curve time was good as compared to other ferrite samples. The sample shows high stability at higher concentration with good sensing response for sensor applications. At lower concentration, europium-doped sample shows low sensing response; however, the low sensing responsible ferrites are used in battery and electronic applications.

DOI: http://dx.doi.org/10.5772/intechopen.90880

Author details

I.C. Sathisha[1], K. Manjunatha[1], V. Jagadeesha Angadi[2*], B. Chethan[3], Y.T. Ravikiran[4], Vinayaka K. Pattar[5], S.O. Manjunatha[6] and Shidaling Matteppanavar[7]

1 Department of Physics, School of Engineering, Presidency University, Bangalore, India

2 Department of Physics, P.C. Jabin Science College, Hubballi, India

3 Department of Physics, Research Centre, JNN College of Engineering, Shimoga, India

4 Department of PG Studies and Research in Physics, Government Science College, Chitradurga, India

5 New Chemistry Unit, Jawaharlal Nehru Centre for Advanced Scientific Research, Bangalore, India

6 Department of Physics, M.S. Ramaiah University of Applied Sciences, Bengaluru, India

7 Department of Physics, Basavaprabhu Kore Arts, Science and Commerce College, Chikodi, India

*Address all correspondence to: jagadeeshbub@gmail.com

References

[1] Jansi Rani B, Saravanakumar B, Ravi G, Ganesh V, Ravichandran S, Yuvakkumar R. Optical and magnetic properties of $CuFe_2O_4$ nanoparticles. Journal of Material Science: Materials in Electronics. 2018;**29**:1975-1984

[2] Srinivasamurthy KM, Angadi VJ, Kubrin SP, Matteppanavar S, Kumar PM, Rudraswamy B. Evidence of enhanced ferromagnetic nature and hyperfine interaction studies of Ce-Sm doped Co-Ni ferrite nanoparticles for microphone applications. Ceramics International. 2018;**44**:18878-18885

[3] Naik CC, Salker AV. Investigation of the effect of fractional In^{3+} ion substitution on the structural, magnetic, and dielectric properties of Co-Cu ferrite. Journal of Physics and Chemistry of Solids. 2019;**133**:151-162

[4] Kurian J, Mathew MJ. Structural, optical and magnetic studies of $CuFe_2O_4$, $MgFe_2O_4$ and $ZnFe_2O_4$ nanoparticles prepared by hydrothermal/solvothermal method. Journal of Magnetism and Magnetic Materials. 2017;**451**:121-130

[5] Manjunatha K, Sathish IC, Kubrin SP, Kozakov AT, Lastovina TA, Nikolskii AV, et al. X-ray photoelectron spectroscopy and low temperature Mössbauer study of Ce^{3+} substituted $MnFe_2O_4$. Journal of Materials Science: Materials in Electronics. 2019;**30**: 10162-10171

[6] Jagadeesha Angadi V, Matteppanavar S, Katti RB, Rudraswamy B, Praveena K. Breaking of ferrimagnetic ordering in Sc^{3+} doped Mn-Zn ferrites due to high energy Gamma irradiation. AIP Conference Proceedings. 2017;**1832**: 130040

[7] Jagadeesha Angadi V, Rudraswamy B, Sadhana K, Ramana Murthy S, Praveen K. Effect of Sm^{3+}–Gd^{3+} on structural, electrical and magnetic properties of Mn-Zn ferrites synthesized via combustion route. Journal of Alloys and Compounds. 2016;**656**:5-12

[8] Othman I, Abu Haija M, Ismail I, Zain JH, Banat F. Preparation and catalytic performance of $CuFe_2O_4$ nanoparticles supported on reduced graphene oxide ($CuFe_2O_4$/rGO) for phenol degradation. Materials Chemistry and Physics. 2019;**238**:121931

[9] Srinivasamurthy KM, Manjunatha K, Sitalo EI, Kubrin SP, Sathish IC, Matteppanavar S, et al. Effect of Ce^{3+} substitution on the structural, morphological, dielectric, and impedance spectroscopic studies of Co–Ni ferrites for automotive applications. Indian Journal of Physics. 2018;**1**:1-12

[10] Galivarapu JK, Kumar D, Banerjee A, Sathe V, Aquilanti G, Rath C. Effect of size reduction on cation distribution and magnetic transitions in $CoCr_2O_4$ multiferroic: EXAFS, magnetic and diffused neutron scattering measurements. RSC Advances. 2016;**6**:63809-63819

[11] Sakthisabarimoorthi A, Martin Britto SA, Dhas R, Jose RM. Influence of erbium doping on the electrical behaviour of $CaCu_3Ti_4O_{12}$ ceramics probed by impedance spectroscopy analysis. Materials Research Bulletin. 2018;**106**:81-92

[12] Oumezzine E, Hcini S, Rhouma FIH, Oumezzine M. Frequency and temperature dependence of conductance, impedance and electrical modulus studies of $Ni_{0.6}Cu_{0.4}Fe_2O_4$ spinel ferrite. Journal of Alloys and Compounds. 2017;**726**:187-194

[13] Sivakumar N, Narayanasamy A, Shinoda K, Chinnasamy CN, Jeyadevan B, et al. Electrical and

magnetic properties of chemically derived nanocrystalline cobalt ferrite. Journal of Applied Physics. 2007;**102**: 013916

[14] Angadi VJ, Choudhury L, Sadhana K, Liu H-L, Sandhya R, Matteppanavar S, et al. Structural, electrical and magnetic properties of Sc^{3+} doped Mn-Zn ferrite nanoparticles. Journal of Magnetism and Magnetic Materials. 2017;**424**:1-11

[15] Chethan B, Raj Prakash HG, Ravikiran YT, Vijayakumari SC, Ramana CHVV, Thomas S, et al. Enhancing humidity sensing performance of polyaniline/water soluble graphene oxide composite. Talanta. 2019;**196**:337-344

[16] Babu Reddy LP, Megha R, Raj Prakash HG, Ravikiran YT, Ramana CHVV, Vijaya Kumari SC, et al. Copper ferrite-yttrium oxide (CFYO) nanocomposite as remarkable humidity sensor. Inorganic Chemistry Communications. 2019;**99**:180-188

[17] Jeseentharani V, George M, Jeyaraj B, Dayalan A, Nagaraja KS. Synthesis of metal ferrite (MFe_2O_4, M = Co, Cu, Mg, Ni, Zn) nanoparticles as humidity sensor materials. Journal of Experimental Nanoscience. 2012;**8**:1-13

[18] Chethan B, Raj Prakash HG, Ravikiran YT, Vijayakumari SC, Thomas S. Polypyyrole based core-shell structured composite based humidity Sensor operable at room temperature. Sensors & Actuators: B. Chemical. 2019; **296**:126639

Chapter 7

Iron Oxides Synthesized in Hypersaline Solutions

Nurit Taitel-Goldman

Abstract

Iron oxides were synthesized in conditions similar to those that prevail in deeps of the Red Sea (2–5M NaCl, temperatures 60–80°C, and pH 6.5–10.4). The main phase that was crystallized was submicron magnetite. Additional phases of feroxyhyte, goethite, and akagenéite were also detected. Magnetite morphology observed through high-resolution scanning electron microscopy (HRSEM) varied between euhedral plates and octahedral or unhedral crystals. The euhedral plates were probably crystallized pseudomorphically after platy green rust or $Fe(OH)_2$ due to its quick crystallization. Size of magnetite varied between 18 and 45 nm. The addition of Si retarded crystal growth, and at Si/Fe = 0.5, short-range ordered phases are formed and not magnetite. This finding is in line with earlier laboratory experiments in which Si was found to retard goethite and lepidocrocite crystallization.

Keywords: magnetite, HRSEM, hypersaline environment, feroxyhyte, goethite lepidocrocite, akagenéite

1. Introduction

Synthesis of iron oxides in hypersaline solutions was performed to imitate conditions that prevail in the Dead Sea and in deeps of the Red Sea. Salinity in the Dead Sea reaches 340 g/l. In the Red Sea, hydrothermal hypersaline brine discharges into the Atlantis II Deep with Cl concentrations in the upper layer of the brine at 67 g of Cl per kg of water (67 g/kg) and at the lower layer of the brine, 158 g/kg. The pH value decreases from 8.13 at RSDW to 5.2 at the lower layer [1]. The temperature of the lower layer is ~68°C measured in 2008 [2]. A narrow channel that connects the southern part of the Atlantis II Deep with the Chain Deeps enables overflowing of the brine of the lower convecting layer to Chain A, B, and C deeps. It was suggested that not only does the overflowing brine feed the adjacent deeps, but a fracture and fissure system enables a sub-bottom flow of the brine of the LCL from the Atlantis II Deep into the Chain and Discovery Deeps, leading to hydrothermal conditions in those deeps [1, 3]. The temperature in Chain A is 52.4–54°C; in Chain B, 46°C; in Chain C, 44.3°C; and in Discovery Deep, 44.7°C [4, 5]. There is no current hydrothermal activity in the Thetis Deep, yet it was active in the past but without a brine that filled the Deep [6].

Iron oxides (hematite-Fe_2O_3 and magnetite-Fe_3O_4), well-crystallized oxyhydroxides (goethite-$\alpha FeOOH$, lepidocrocite-$\gamma FeOOH$, feroxyhyte-$\delta FeOOH$, and akaganéite-$\beta FeOOH$), and short-range ordered oxyhydroxides (ferrihydrite-$Fe_5HO_8{\cdot}4H_2O$ and singerite-$SiFe_4O_6(OH)_4H_2O$) precipitated in the hypersaline environment of the Dead

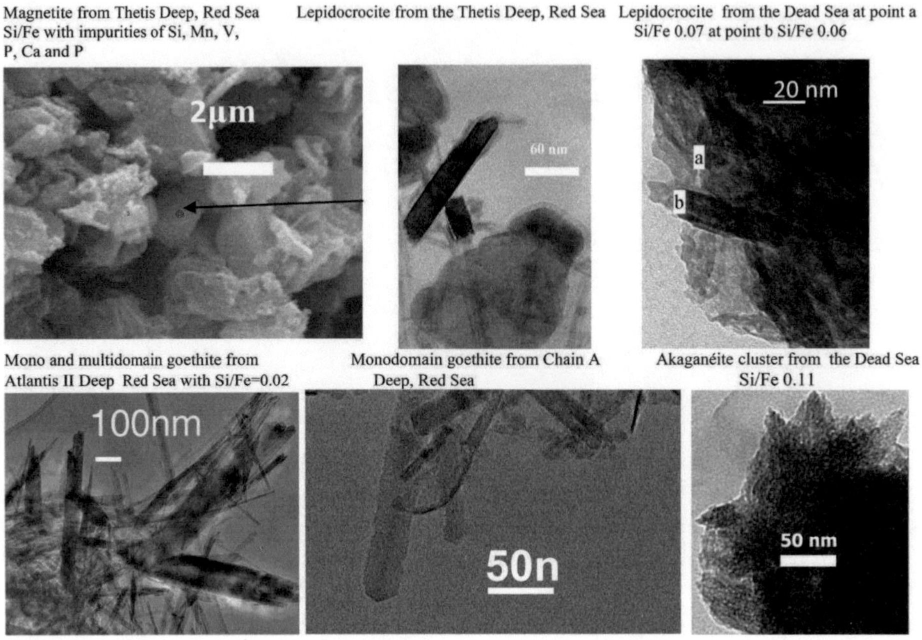

Figure 1.
Pictures obtained by electron microscopy: Magnetite from the Thetis deep (scanning electron microscope (SEM)); lepidocrocite from the Thetis deep; lepidocrocite from the Dead Sea; mono- and multi-domain goethite from Atlantis II deep, Red Sea; mono-domain goethite from chain a deep, Red Sea; and cluster of akaganéite from the Dead Sea (high-resolution transmission electron microscope).

Sea and in the deeps of the Red Sea (Atlantis II, Chain A, Chain B, Discovery, and Thetis) [7–10] (**Figure 1**). Precipitation of these phases occurs at hypersaline environment with elevated temperatures and varying pH. Morphology of the crystals and their size is observed by high-resolution transmission electron microscopy (HRTEM). Using electron diffraction enables identifying the crystallographic structure. Goethite and lepidocrocite have orthorhombic structure, akaganéite has monoclinic structure, and feroxyhyte has hexagonal structure [11].

Short-range ordered ferrihydrite recrystallizes into well-crystallized phase, yet its initial morphology of plates at the size of <10 nm is preserved. Singerite (100 nm) disintegrates into iron-rich clay mineral [12]. Magnetite has a cubic morphology of octahedron or cubes 2–4 μm as observed in the Thetis Deep of the Red Sea.

Goethite has acicular form and changes at elevated pH from mono-domain to multi-domain rods. Crystal size at the Dead Sea area and the Red Sea varied from few 100 nm to 3 μm.

Lepidocrocite usually precipitates at fast oxidation rate at the presence of chloride, and its morphology changes from plates formed at lower pH to multi-domain rods that are formed at elevated pH. Lepidocrocite crystals size varied from 100 nm in the Dead Sea to 300 nm in the Red Sea deeps.

Rods and multi-domain akaganéite were observed at the Dead Sea with crystal size that varied from 100 nm to 1 μm.

Feroxyhyte plates are formed at very high oxidation rate and had a plate morphology with crystals size that reached 300 nm.

In earlier study, synthesis was performed imitating the upper convecting layer of the Atlantis II Deep with lower temperatures and salinity [13]. In this research, iron oxides were synthesized at higher pH, elevated temperatures, and hypersaline brines.

2. Methods

Iron oxides were synthesized under changing conditions of salinity, temperature, pH, and oxidation rates.

NaCl salt (Loba Chemie) was used to prepare NaCl solutions (2, 4, and 5M) that were used as a matrix and were kept in water bath at 40, 60, 70, and 80°C. Prior to synthesis, N_2 was bubbled through the solutions for 20 minutes to remove dissolved oxygen. pH buffering of the solutions was accomplished by adding either NaOH (Daejung) or $NaHCO_3$ (Carlo Erba) in small amounts, hence slightly changing Na concentration. pH of the solutions were 5.5, 7, 8.2, and 10.5. Cl concentration also changed due to dissolution of $FeCl_2{\cdot}4H_2O$ (Sigma-Aldrich) that was chosen for the Fe^{2+} solutions to yield a concentration of 0.06M.

Fe oxidation was carried out by introducing air at flow rates of 25, 40, 110, and 200 ml/min which was monitored with a flow meter and was kept stable during the hours of synthesis.

In order to isolate Si effects on the crystallization of iron oxides, Na_2SiO_3 (Sigma) was added to some of the solutions. To avoid any side effects, all of the samples were synthesized using the same polyethylene ware and under constant stirring speed. The precipitates were slightly washed and freeze-dried immediately after their synthesis.

Analyses of the precipitates were performed using X-ray diffraction (XRD).

The fitting of the peak profiles was performed by using Pseudo-Voigt function with the APD computer program developed by Philips Export B.V. A least square process using "Celsiz" software did the unit cell refinement.

Sample morphology was observed through transmission electron microscopy (TEM) and high-resolution scanning electron microscopy (HRSEM). Transmission electron microscopy was carried out on a JEOL FasTEM 2010 electron microscope equipped with the Noran energy dispersive spectrometer (EDS) for microprobe elemental analyses.

All chemical analyses were obtained by point analyses with beam width of 25 nm and are presented as atomic ratios. A NORAN Standardless Metallurgical Thin Films program based on the *Cliff-Lorimer* ratio technique with an accuracy of about 5% was used for calculations. CuK_α line was used for spectrometer calibration. Crystalline phases were identified using selected area electron diffraction (SAED) in the TEM.

3. Results and discussion

Initial identification of iron oxides was obtained by using X-ray diffractions (**Figure 2**). X-ray diffraction of all samples synthesized yielded the following results: At elevated temperatures, alkaline media and concentrated brine magnetite was the main phase that precipitated. At neutral to slightly acidic conditions, lepidocrocite, akaganéite, and goethite co-precipitated. Additional $NaHCO_3$ caused precipitation of siderite at 60°C.

In this study, pictures of iron oxides obtained by high-resolution scanning electron microscope and high-resolution transmission electron microscope are presented.

3.1 Magnetite FeO-Fe_2O_3

Unit cell parameters of magnetite vary from 0.8373 to 0.8396 nm with upper limit of 0.8396 nm, identical to that of well-crystallized magnetite [13]. Crystallite size changes from 18 to 45 nm. The smallest crystallites were obtained at pH 7.5 in 4 or 5M NaCl at all temperatures, whereas the largest crystallites were obtained at under elevated pH (9.5) and temperatures. The lowermost unit cell parameter was obtained for samples that crystallized at 60°C pH 8.5 and in 4M NaCl matrix. The largest unit

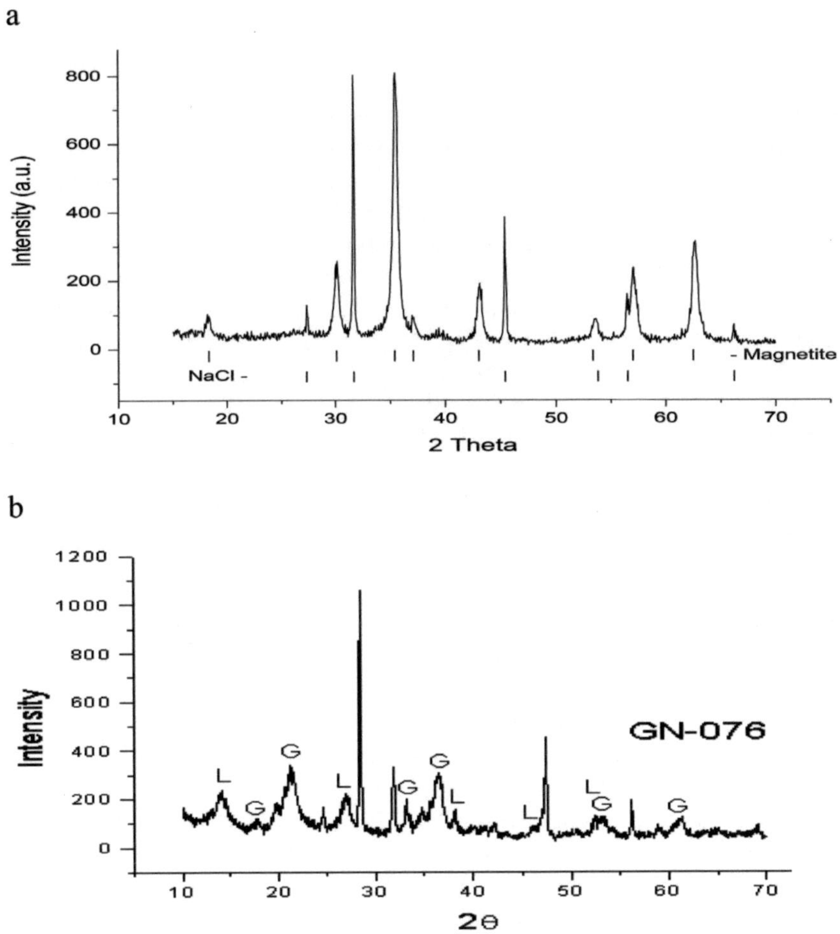

Figure 2.
X-ray diffraction patterns of (a) synthesized magnetite at 70°C, 5M NaCl, and pH 7.5; (b) goethite and lepidocrocite in sample from the Atlantis II deep, Red Sea.

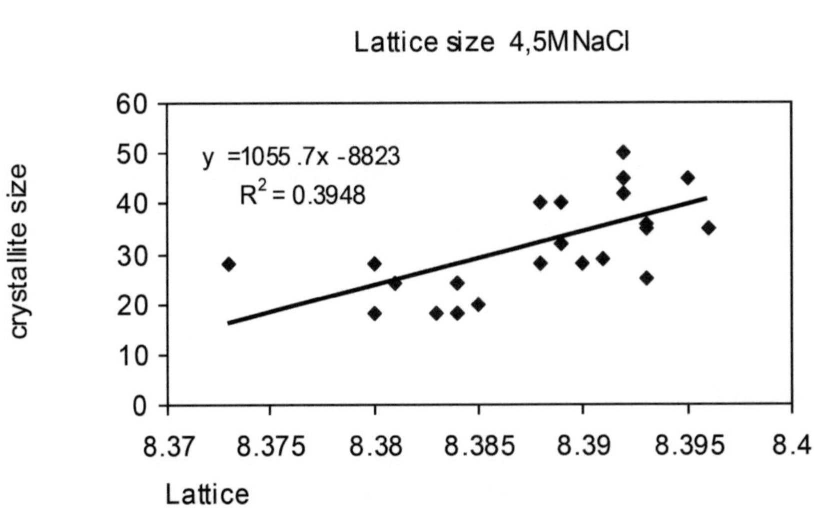

Figure 3.
Lattice parameter (nm) and crystalline sizes of magnetite precipitated in brines with concentrations of 4 and 5M NaCl.

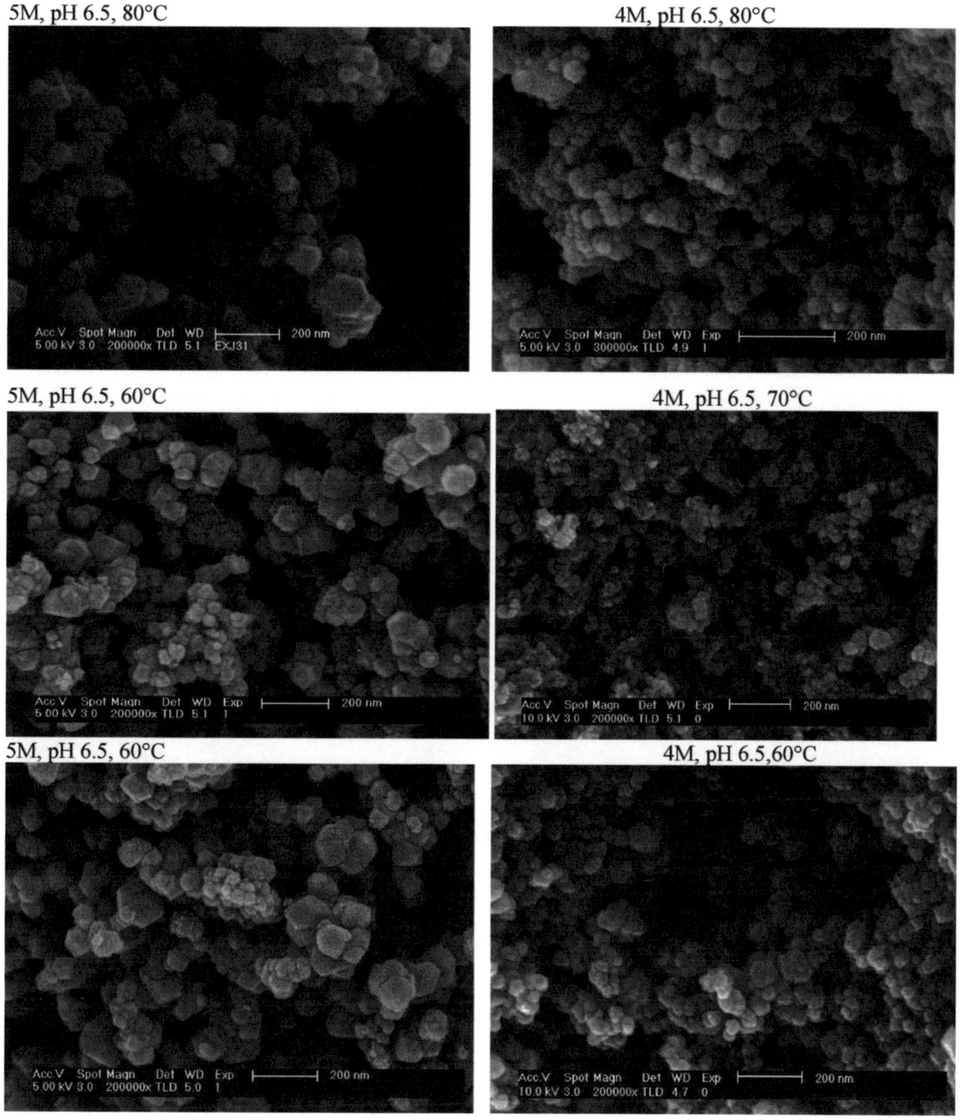

Figure 4.
Pictures obtained by HRSEM presenting the morphology of magnetite crystals precipitated at salinity of 4 and 5M NaCl, pH 6.5, and temperature of 60–80°C.

cell parameter was obtained for samples that crystallized under elevated pH 9.5, 5M NaCl and at 80°C. Overall, crystallite size and unit cell parameters were mainly affected by pH values, rather than by salinity or temperature changes (**Figure 3**).

Morphology of the crystals formed varied between unhedral to unhedral-octahedral, cubes, and hexagonal plates. The morphology of the crystals mainly results from the pH of the brines in **Figures 4–8**.

Magnetite Fe_3O_4: FeO-Fe_2O_3 crystallizes in cubic system with a = 0.8396 nm.

Crystallization from Fe^{2+} solutions usually involves crystallization of hexagonal flakes of $Fe(OH)_2$, which transforms to magnetite in moderately alkaline solutions (pH > 8). Under slightly acid to slightly alkaline conditions, green rust phases are formed, and upon further oxidation they are transformed into goethite and/or lepidocrocite [11]. Formation of magnetite in the hypersaline brines results from slow oxidation, elevated temperatures, and higher pH. The morphology of the magnetite

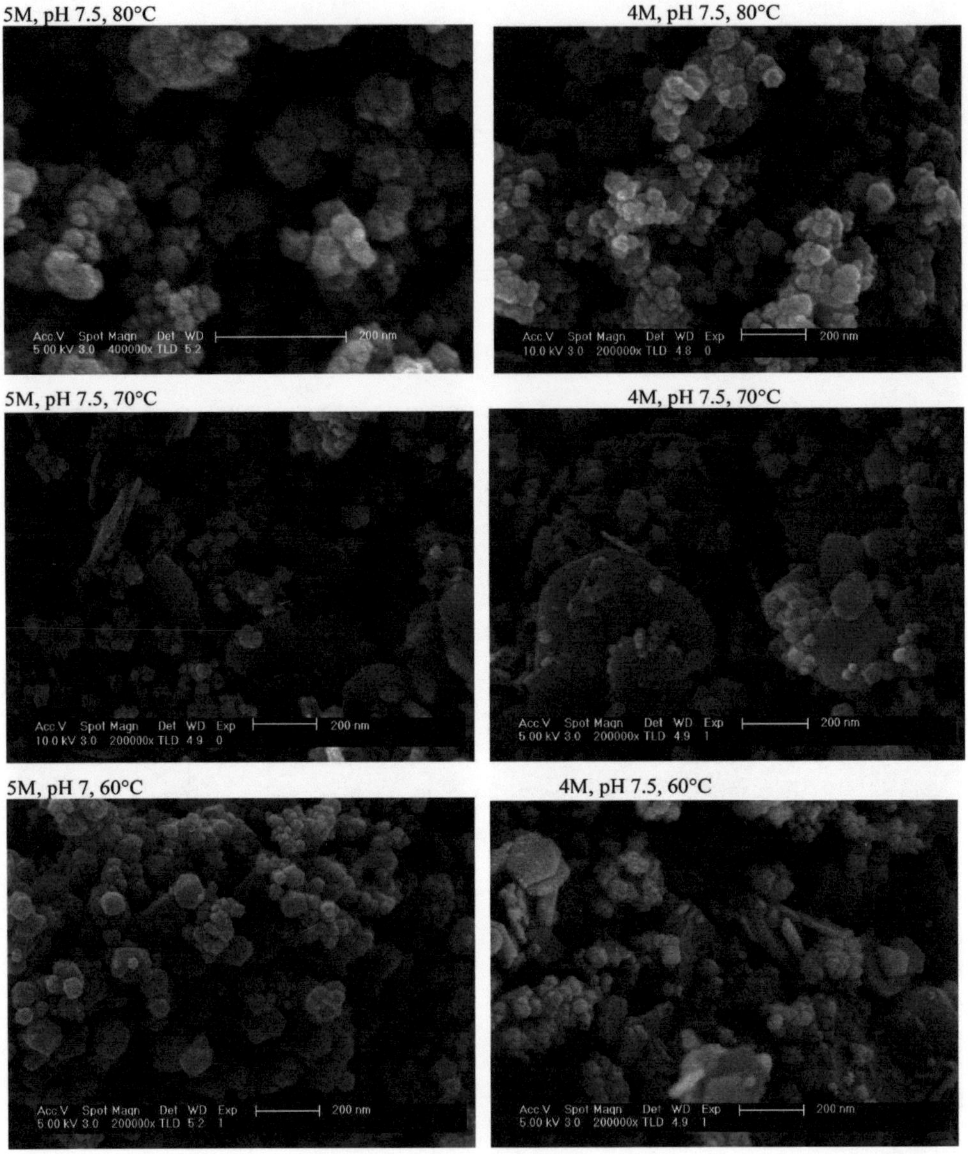

Figure 5.
Pictures obtained by HRSEM presenting the morphology of magnetite crystals precipitated at salinity of 4 and 5M NaCl, pH 7.5, and temperature of 60–80°C.

crystals was mainly affected by pH. In strongly alkaline brine, hexagonal flakes of $Fe(OH)_2$ were initially formed recrystallized into magnetite, preserving the initial morphology.

The morphology of magnetite of the Thetis Deep indicates that precipitation occurred at lower pH and lower salinity, suggesting that no brine pool filled the Thetis Deep [6].

3.2 Goethite (α-FeOOH)

Goethite precipitated along with other iron oxides. The morphologies of the crystals formed varied between mono-domain and multi-domain crystals. Twinning and star-shaped multi-domain crystals were formed at elevated temperatures. In higher salinity of 5M NaCl, goethite precipitated at pH 8.2 and 40°C, and at pH 7 and

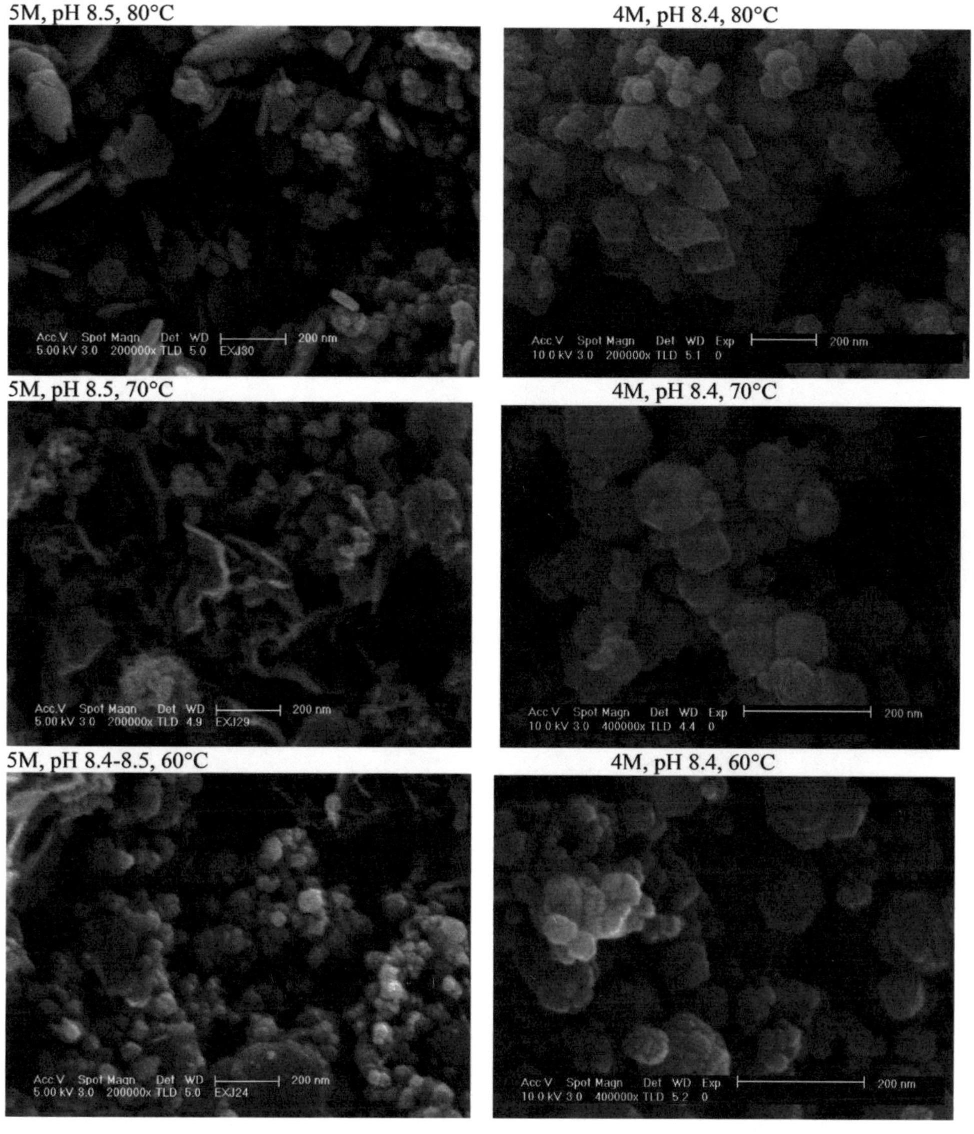

Figure 6.
Pictures obtained by HRSEM presenting the morphology of magnetite crystals precipitated at salinity of 4 and 5M NaCl, pH 8.5, and temperature of 60–80°C.

40°C, it co-precipitated with lepidocrocite and akaganéite. Adding $NaHCO_3$ to the solution caused co-precipitation of goethite with lepidocrocite at 40°C in 2M NaCl, co-precipitation of goethite with magnetite at 60°C and at 5M NaCl solutions, at pH 8.2, only magnetite precipitated magnetite [13] (**Figure 9**).

3.3 Lepidocrocite (γ-FeOOH)

Lepidocrocite precipitated at lower temperatures than magnetite. By using NaCl solutions of 5M and temperature of 40°C, the morphology of the crystals formed was affected by the pH. Plates crystallized at pH 5.5, rods at pH 7, and multi-domain crystals at pH 8.2 (**Figure 10**). The difference between the morphologies results from faster crystallization of 010 along c direction in crystals formed at higher pH leading to multi-domain crystals at pH 8.2. Similar results were also obtained at 25°C [13].

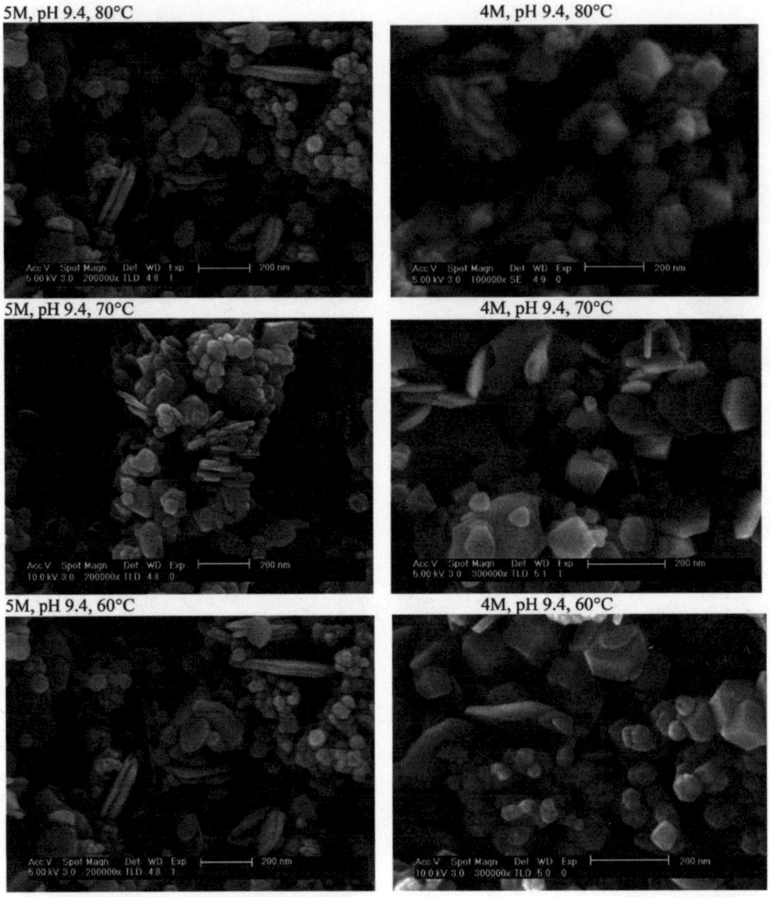

Figure 7.
Pictures obtained by HRSEM presenting the morphology of magnetite crystals precipitated at salinity of 4 and 5M NaCl, pH 9.4, and temperature of 60–80°C.

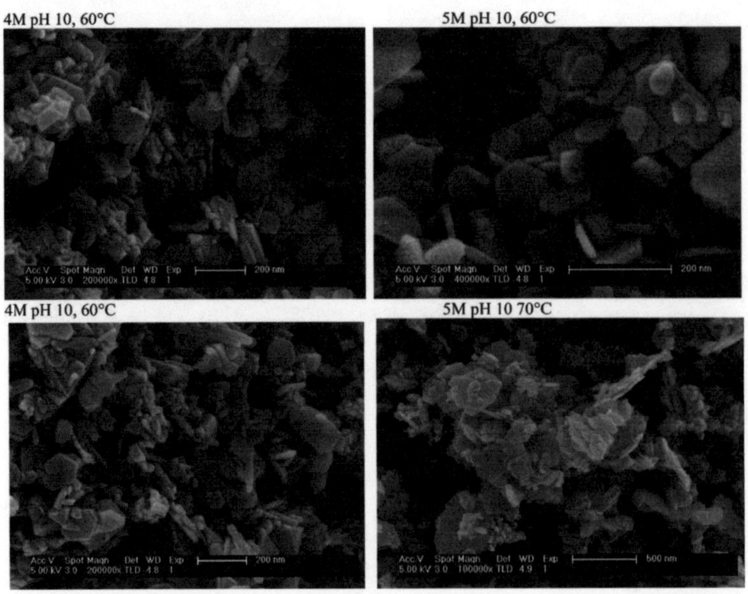

Figure 8.
Pictures obtained by HRSEM presenting the morphology of magnetite crystals precipitated at salinity of 4 and 5M NaCl, pH 10, and temperature of 60–70°C.

Figure 9.
Goethite crystals observed through HRTEM. The crystals were synthesized at 40°C, in solutions of 2 and 5M NaCl and with additional $NaHCO_3$ or at pH 8.2.

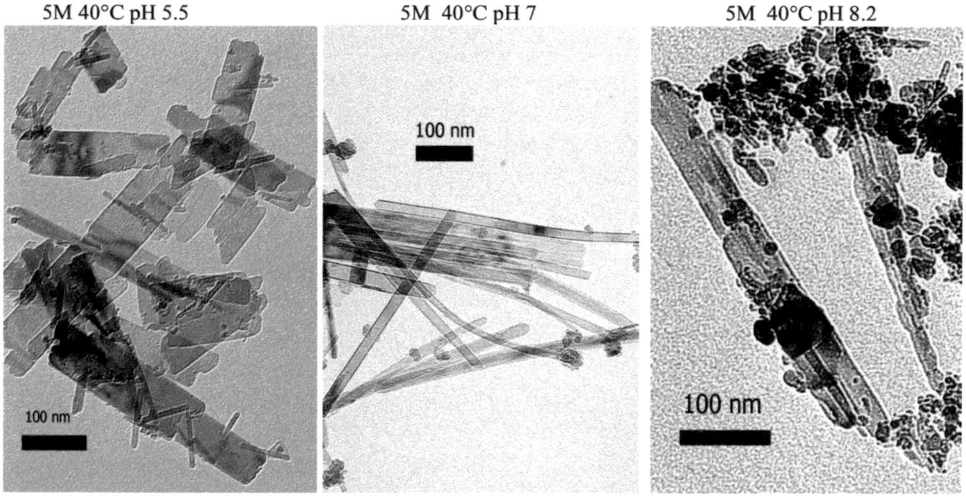

Figure 10.
Lepidocrocite crystals observed through TEM. Crystals were formed in NaCl solutions of 5M concentration at 40°C and pH 5.5, 7, and 8.2. At pH 8.2 magnetite crystals co-precipitated with multi-domain lepidocrocite.

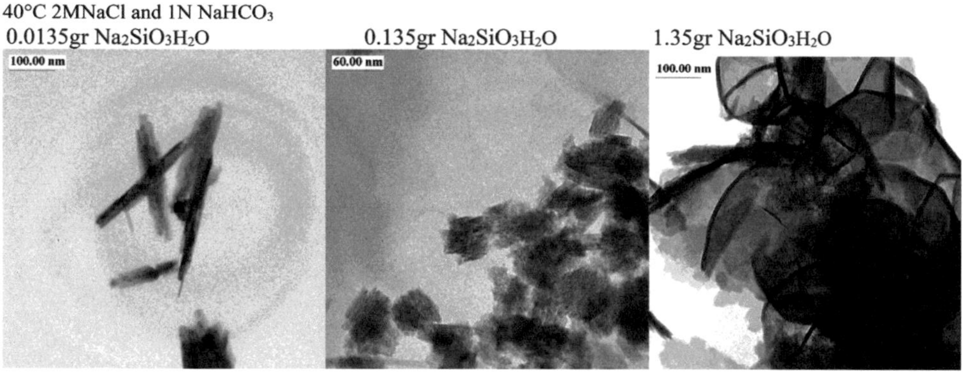

Figure 11.
Pictures obtained by HRTEM presenting the effect of additional $Na_2SiO_3H_2O$ to solution of 4M NaCl, at 40°C.

3.4 Effect of additional Si

Addition of $Na_2SiO_3H_2O$ into the solutions was selected to see the effect of Si on iron oxide crystallization. In precipitation of magnetite crystals, $Na_2SiO_3H_2O$ was added to a solution of 4M NaCl, pH 7.5 at a temperature of 60°C. The presence of Si affected crystal size, which became smaller, from 30 to 26 nm, and unit cell parameter decreased from 0.8392 to 0.8380 nm.

Addition of 0.0135 g $Na_2SiO_3H_2O$ to the solutions at the initial crystallization stage hindered crystal growth of goethite crystals (**Figure 11**). Elevated amounts of Si added as 0.135 g Na_2SiO_3 and 1.35 g caused formation of short-range ordered phases (suggested name singerite) [12].

4. Conclusions

Magnetite crystalized in hypersaline 2–5M brine at elevated temperatures 60–80°C and pH 6.5–10. At high alkalinity, their initial formation stage of hexagonal plates was preserved. At lower alkalinity, their morphology was unhedral cubes. Additional Si to the brines caused smaller crystals.

Lepidocrocite and goethite were formed at lower temperatures (40°C). High alkalinity effected crystal morphologies causing multi-domain structure in goethite and in lepidocrocite and twinning formation.

Additional Si hindered crystal growth.

Author details

Nurit Taitel-Goldman
The Open University of Israel, Raanana, Israel

*Address all correspondence to: nuritg@openu.ac.il

References

[1] Hartmann M, Scholten JC, Stoffers P, Wehner F. Hydrographic structure of brine-filled deeps in the Red Sea—New results from the Shaban, Kerbit, Atlantis II, and discovery deep. Marine Geology. 1998a;**144**:311-330

[2] Swift A, Bower AS, Schmitt R. Vertical, horizontal, and temporal changes in temperature in the Atlantis II and discovery hot brine pools, Red Sea. Deep Sea Research Part I: Oceanographic Research Papers. 2012;**1**(64):118-128

[3] Pierret MC, Clauer N, Bosch D, Blanc C, France-Lanord C. Chemical and isotopic (87Sr/86Sr, d18O, dD) constraints in the formation processes of Red Sea brines. Geochimica et Cosmochimica Acta. 2001;**65**:1259-1275

[4] Ramboz C, Danis M. Superheating in the Red Sea? The heat-mass balance of the Atlantis II deep revisited. Earth and Planetary Science Letters. 1990;**97**:190-210

[5] Hartmann M, Scholten JC, Stoffers P. Hydrographic structure of brine-filled deeps in the Red Sea: Correction of Atlantis II deep temperatures. Marine Geology. 1998b;**144**:331-332

[6] Pierret MC, Clauer N, Bosch D, Blanc G. Formation of Thetis deep metal-rich sediments in the absence of brines, Red Sea. Journal of Geochemical Exploration. 2010;**104**:12-26

[7] Taitel-Goldman N, Bender-Koch C, Singer A. Lepidocrocite in hydrothermal sediments of the Atlantis II and Thetis deeps, Red Sea. Clays and Clay Minerals. 2002;**50**:186-197

[8] Taitel-Goldman N, Bender-Koch C, Singer A. Si-associated goethite in hydrothermal sediments of the Atlantis II and Thetis deeps, Red Sea. Clays and Clay Minerals. 2004;**52**:115-129

[9] Taitel-Goldman N, Ezerski V, Mogilyanski D. High resolution transmission electron microscopy study of Fe-Mn oxides in the hydrothermal sediments of the Red Sea deeps system. Clays and Clay Minerals. 2009;**57**:465-475

[10] Taitel-Goldman N, Ezerski V, Mogilyanski D. Nano-sized iron-oxides in the Dead Sea area. Journal of Earth Science. 2016;**2**:94-104

[11] Cornell RM, Schwertmann U. The Iron Oxides: Structure, Properties, Reactions, Occurrences. Weinheim, Germany: Wiley VCH; 2003

[12] Taitel-Goldman N, Singer A. Metastable Si-Fe phases in hydrothermal sediments of Atlantis II deep, Red Sea. Clay Minerals. 2002;**37**:235-248

[13] Taitel-Goldman N, Singer A. Synthesis of clay-sized iron oxides under marine hydrothermal conditions. Clay Minerals. 2002;**37**:719-731

Chapter 8

Preparation and Characterization of Fe_2O_3-SiO_2 Nanocomposite for Biomedical Application

Violeta N. Nikolić

Abstract

The scope of this chapter is to get deeper insight into the correlation between synthesis parameters and magnetic behavior of the nanocomposite materials containing hematite (α-Fe_2O_3) nanoparticles. Potential applications of nano-hematite in biomedicine are listed in the short overview. Then, basic requirements necessary for synthesis of high-quality nanoparticles for biomedical application are summarized. The next part of the chapter is devoted to the sol-gel synthesis that is recognized as suitable for preparation of the nanocomposite materials containing α-Fe_2O_3 nanoparticles. Having in mind that sol-gel method considers preparation of hematite nanoparticles via Fe_2O_3 phase transformations initiated by thermal treatment at high temperatures, coexistence of the other iron oxides (such as ε-Fe_2O_3) with α-Fe_2O_3 phase is commented. Special attention is paid on mechanism of the critical field (which is in literature usually denoted as coercivity field) alterations. Diffraction patterns and hysteresis measurements of the chosen samples containing hematite nanoparticles in the silica matrix are represented. Finally, variations in the observed measured critical field values are discussed.

Keywords: nano α-Fe_2O_3-phase, nano ε-Fe_2O_3-phase, silica, synthesis parameters, intrinsic coercivity field, biomedical application

1. Introduction

Hematite (α-Fe_2O_3) has been thoroughly investigated during the centuries, since it is one of the most abundant minerals in the earth's crust: on the surface as well as at the bottom of the sea [1, 2]. From the beginning of its discovery to the present days, α-Fe_2O_3 gained attention of the scientific community due to its magneto-structural properties, high resistance to corrosion, easy accessibility, wide distribution in natural environment, and biocompatibility. Accordingly, α-Fe_2O_3 is recognized as a material of significance in different scientific areas [3–12].

The usage of hematite in conventional biomedicine has been enabled by the breakthrough and development of the nanoscience. Progress in this field is achieved by detailed research of iron oxide polymorphs' physical and chemical properties on the nanoscale. Properties of nm-sized particles are significantly differed from their bulk counterparts due to various nano-size-related effects, appeared as a consequence of the changed ratio of surface and volume atoms. Approaching nanometer dimensions, the ratio of the surface atoms in the overall nanoparticle volume drastically increased by decreasing particle size, favoring the role of the surface effects in the characteristic behavior of the nanoparticles.

Although the most attention in this chapter will be paid to magnetic properties of hematite nanoparticles, it is important to mention that overall behavior of nano-hematite is characterized by its electromagnetic response which determined biomedical application of nanoparticles. Due to coupling between nano-hematite electrical and magnetic fields, it is clear that electrical properties are contributing to the final hematite application as well as magnetic properties. Noteworthy, difference between conductivity of bulk and nano-hematite widespread biomedical application of this iron oxide polymorph application. According to the analysis of the density of electronic states, the bulk hematite showed the charge-transfer insulator nature [13]. Differed from bulk hematite, nano-hematite is an n-type semiconductor, with a bandgap of ~2.2 eV [14]. Consequently, the increase in conductivity enables the extended application of nano-hematite in different biomedical areas, such as transfer of electrical signals in biosensors, tissue engineering, neural probes, drug delivery, or diagnosis, and therapy of human diseases [15].

On the other hand, transition from bulk to nano dimensions resulted in the significant change of its magnetic behavior. Magnetic behavior of bulk hematite is determined by Neel temperature (T_N) (~950 K) and Morin temperature (T_M) (~260 K) that represent the temperatures upon which the hematite magnetic ordering is changed. Above T_N, hematite is characterized by paramagnetic structure. In the temperature range between T_N and T_M, hematite showed a weak ferromagnetic ordering, while under T_M it is antiferromagnetic [16]. Hematite magnetic configurations are defined by the magnetic interactions (magnetic ordering of bulk materials is mostly influenced by the exchange interaction) [17].

Magnetic structure of bulk hematite is represented by different regions of a macroscopic system broken symmetry in different ways, the so-called domains. Domains present small regions within each of which the local magnetization achieves the saturation value [17], while interface between the adjacent regions presents the domain wall. The processes of magnetization and demagnetization of materials occurred through the movement of the domain walls and change of the domain boundaries, consequently bringing to the increase/decrease of a domain size.

From the aspect of magnetic interactions, decrease of the particle size revealed a dominant role of dipole-dipole interactions in the ordering of nanoparticles' magnetic moments, which is negligible in the case of bulk hematite, since magnetic moment of bulk magnetic material is significantly lower than the moment of the nanoparticle [17]. To get a better insight in the changes in the strength ratio of bulk and nano-magnetic interactions that are responsible for maintaining of long-term ordering of magnetic moments, it is important to understand the origin of the increase in the value of nanoparticle magnetic moment. The increase of magnetic moment occurred as a consequence of lowering dimensionality of bulk materials and has been explained by the absence of multi-domain structure and appearance of single-domain nanoparticle structure. The balance between the anisotropy and exchange energies is required for formation of finite-size domain walls [18], resulting in the presence of some critical diameter size, below which nanoparticle is single-domain [18], Eq. (1):

$$r_c \approx 9\left[(J_{ab} \cdot K_a)^{1/2} / \mu_o \cdot M_s^{\,2}\right] \tag{1}$$

where J_{ab} is the exchange integral, K_a is the anisotropy constant, and M_s is the saturation magnetization.

Noteworthy, macroscopic magnetic properties of nanoparticles are mostly influenced by Zeeman energy, thermal energy, and anisotropy energy. The relations of mentioned energies [19] enable the appearance of a new magnetic state in nanomaterials, characterized by a presence of single-domain particles—superparamagnetic nanoparticles. The main characteristic of superparamagnetic nanoparticle system is the absence of coercivity and remanence at room temperature, which enables the application of magnetic properties of nano-hematite in biomedicine.

Considering the fact that nanomaterials possesses high surface-to-volume ratio and increased surface activity [11], it is obvious that hematite nanoparticles show various magnetic behavior dependent on the size and shape of the particles [12, 20]. Also, it is important to notice that observation of dependence of magnetic or electric properties of nano-hematite on synthesis conditions is enabled by the change of particle size, carrier density (that is dependent on the particle size), domain size, and structure of the synthesized nanoparticles. Furthermore, alteration of synthesis conditions enabled tailoring of their magneto-structural properties and a variety of new applications.

When we are dealing with potential usage of nano-hematite particles in biomedicine, it should be emphasized that biomedical application requires utilization of nanocomposite materials. For preparation of nanocomposites containing α-Fe_2O_3 nanoparticles convenient for this purpose, silica is recognized as a suitable material.

In order to get deeper insight in the magnetic behavior of the nanocomposites, samples that contained hematite nanoparticles in silica matrix are often prepared by sol-gel method that involves formation of hematite nanoparticles by the phase transformations of the Fe_2O_3 polymorph (maghemite (γ-Fe_2O_3) and epsilon phases (ε-Fe_2O_3)). Due to the presence of particle size distribution in the nanomaterials, special attention should be addressed to the problem of coexistence of different iron oxide species within the silica matrix.

Usually the goal of the synthesis is the preparation of the samples characterized by high purity, containing only one iron oxide phase. In some cases the usage of precisely one phase of the iron oxide polymorph is not of crucial importance. For example, magnetite and maghemite nanoparticles are characterized by similar magneto-structural properties and thus could be used together for the preparation of the magnetic ferrofluids. Noteworthy, synthesis of the nanocomposite materials containing this type of iron oxide nanoparticles is so common in literature that the scientific community accepted the abbreviation “SPION” (superparamagnetic iron oxide nanoparticles) to describe spinel iron oxide species [21, 22].

Comparing the magnetic behavior of hematite nanoparticles with the magnetic properties of the other iron oxides (spinel or epsilon phase), it is certain that α-Fe_2O_3 cannot be used together with spinel phases due to very different magnetic properties. On the other side, at the moment it is not possible to claim with certainty whether it could be used together with the epsilon phase or not. This is a consequence of the insufficient knowledge about magnetic properties of these two iron oxide phases. Literature data revealed that pure hematite nanoparticles are characterized by the intrinsic coercivity field (H_{ci}) value of 1.7 kOe [23], although nano-sized α-Fe_2O_3 in silica matrix can achieve coercivity of 4.3 kOe [24]. Under the certain size limit, hematite nanoparticles showed superparamagnetic (SPM) behavior [12]. The presence of the other ions significantly alters hematite coercivity. Even bulk hematite doped with alumina ions reached coercivity of >8 kOe [25]. It is similar with the lack of the knowledge regarding epsilon H_c value: dependent on the synthesis conditions, epsilon nanoparticles showed different coercivity. This phase is characterized by high room temperature coercivity (10–20 kOe) [26, 27]. Nevertheless, some literature reports depicted the lowered epsilon H_c value (8 or 2.4 kOe [28, 29]). As well, epsilon nanoparticles could be prepared in order to display SPM behavior [30]. Although hematite nanoparticles cannot achieve coercivity of 10–20 kOe, there is a certain interval of H_{ci} values during which the hematite and epsilon phase coercivities could overlap. Likewise, it is important to point out that coercivity of the samples containing both phases, hematite and epsilon, significantly varies dependent on the synthesis conditions.

The aim of this chapter was to examine in more detail the correlation between synthesis parameters and magnetic properties of nanocomposites containing pure hematite phase or hematite phase in combination with the SPM epsilon phase. A better insight in the measured magnetic field (which is in literature usually denoted as coercivity field) variations dependence on the synthesis conditions is of importance for improvement of the current efforts in understanding of the magnetic properties of hematite phase. Also, some difficulties inherent in studying influence of the variation of synthesis conditions onto the magnetic behavior of the examined samples are highlighted. Results summarized in this chapter could facilitate application of nano-hematite in biomedicine.

2. Overview of nano-hematite applications in biomedicine

Plenty of synthesis pathways for production of the nano-hematite enabled formation of hematite nanoparticles characterized by different properties, which determine their application. There are a lot of reasons for a biomedical application of nano-sized α-Fe_2O_3: low cost, long-term chemical stability, and nontoxicity. Up until today, nano-hematite is mostly used as a starting material for preparation of multifunctional nanocomposite particles that found application in different areas of biomedicine. In order to obtain appropriate candidate for biomedical application, nanocomposite materials containing hematite nanoparticles are prepared by a few steps of synthesis procedures.

Some of the biomedical applications of α-Fe_2O_3 nanoparticles are listed below.

Nano-hematite could be used as a starting material for the synthesis of platforms, presenting promising functional nanomaterials for drug delivery and hyperthermia treatments. Liu et al. synthesized α-Fe_2O_3 nanoparticles by hydrothermal method [31]. Particles were further coated with a nonporous silica (Fe_2O_3@SiO_2) and subsequently treated with an organosilicate-incorporated silica by simultaneous sol-gel polymerization of tetraethoxysilane (TEOS) and n-octadecyltrimethoxysilane (C_{18}TMS). Final step of the synthesis considered reduction of the hematite cores to magnetite. Obtained nanocomposite platforms are used as smart-targeted drug delivery materials for further in vivo evaluation of cancer therapies [31]. Another application of the platforms based on the usage of nano-hematite as a starting material is considered a preparation of asymmetric hematite-silica nanocomposites (JFSNs) as multifunctional peroxidase mimetics that found application in glucose colorimetric biosensing [32].

On the other hand, a combination of mesoporous nano-hematite with carbon quantum dots enabled preparation of the nanomaterial that showed promising properties for the application in visible photo-light photocatalysis [33]. Due to very good photocatalytic properties, excellent biocompatibility, and high chemical stability, carbon quantum dots/mesoporous hematite nanocomposites could be used in numerous biomedical applications, such as photodynamic therapy for cancer treatment, drug delivery systems, cell imaging, biosensors for biological assay, and genetic engineering [34].

Mirzaei et al. investigated the usage of materials consisting nano-hematite in biosensor technologies [35]. Nanocomposite material was prepared by Pechini sol-gel method that involved the formation of a complex between hematite nanoparticles and citric acid, followed by an esterification reaction with ethylene glycol. Since hematite nanoparticles are displaying good electrical and sensing stability, nanocomposite material is used as a highly stable and selective biochemical sensor for detection of ethanol and monitoring alcohol consumption [35, 36]. Another biosensing application of hematite nanoparticles denoted the application of anodization method that enables synthesis of highly ordered hematite nanotube array on a patterned SiO_2/Si substrate. Prepared nanomaterial showed an excellent selectivity and ppb-level detection limits toward acetone, depicting its promising application for breath analyzers to diagnose diabetes mellitus [37]. As well, nano-hematite is recognized as a suitable material for magnetically assisted binding assays (measurement of the concentration or potency of a substance by its effect on living cells or tissues) performed by using magnetically labeled binding members [38].

It is important to notice that SPM hematite nanoparticles also could be utilized for biomedical applications, in fabrication of biomolecular sensor system, used for detection of intravenously introduced nanoparticles. Litvinov showed that α-Fe_2O_3 nanoparticles could be used as magnetoresistive nanosensors designed for sensing biomolecule-conjugated nanoparticles (different targets could be detected, such as cell surface receptor, protein, nucleic acid, mRNA, genomic DNA, etc.) [37].

Recent scientific work on *α-Fe_2O_3* revealed a potential application of hematite nanoparticles in genotyping, since results of scientific investigation confirmed the presence of interaction between appreciably high concentration of hematite nanoparticles and drying pattern of a sessile droplet of genomic deoxyribonucleic acid (DNA) [38].

3. Prerequisite conditions for preparation of high-quality nano-hematite particles

An important step in the usage of materials that contained nano-hematite in biomedical application presents synthesis of high-quality nano-hematite particles and high control of its magnetic behavior. To use hematite nanoparticles in biomedical purposes, it is necessary first to modify the surface of nano-hematite.

The nanoparticle surface presents a key factor that determines biocompatibility and enables cell adhesion of the particle injected in the human body. Accordingly, the surface of the nano-hematite particles, predetermined for biomedical application, has to meet few basic requirements:

1. Biocompatibility: non-toxicity for human organism is prerequisite for the application in biomedicine.

2. Monodispersity: uniform nanoparticle size and shape minimized interparticle interactions and agglomeration. This task is not completely overcome up until today, due to the presence of the particle size and shape distribution. For that reason, different synthesis strategies are employed with the aim to improve the knowledge regarding achieving nanoparticles' monodispersity [39, 40]. Reaching monodispersity would allow improvement of the control of magnetic behavior of the overall sample.

3. Functionalization: particles should possess high efficiency for binding target molecules, and non-specific binding should be avoided. In order to ensure the

listed requirements, hematite nanoparticles for application in biomedicine are coated by biocompatible materials (usually with silica, although different materials, such as dextran or citric acid, also could be applied [41–44]) and then further functionalized by attaching groups on the surface (different antibodies, oligonucleotides, or peptide ligands, depending on the desired application [45–47]) via various chemical methods [48–52]. An alternative way for production of suitable nanocomposites presents performing core-shell strategy or encapsulation of the particles in a silica matrix.

4. Sol-gel synthesis

In order to better examine the magnetic behavior of the hematite nanoparticles that could be used as a starting material for different biomedical applications, nanocomposite materials based on hematite are often prepared by sol-gel method. From the point of view of correlation of the synthesis conditions with characteristics of the investigated nanomaterial, this synthesis method is of great significance.

Advantages of this type of synthesis are low price of the chemicals, gelation process under ambient conditions, as well as possibility of the synthesis of very small nanoparticles (~1 nm) [53]. Basic compounds used in the sol-gel synthesis are iron ion precursor, silica ion precursor (tetraethyl orthosilicate, TEOS, or tetramethyl orthosilicate, TMOS), water, and the compound miscible with (mutually nonmiscible) alkoxide precursor and water (ethanol or methanol, depending on the usage of TEOS or TMOS, respectively).

Mechanism of the sol-gel synthesis contains few stages. The first stage consisted nucleation of the Fe_2O_3 and SiO_2 nanoparticles during the hydrolysis of TEOS. Reactions of condensation and polycondensation occurred during the mixing solution and resulted in the nanoparticle growth through the process of Ostwald ripening [53]. Mentioned processes occur at room temperature, conditioning the usage of catalyst, which initiates the changes in the structure and properties of the resulting material. The aging of the prepared sol enables its conversion into gel, which presents the second important stage, followed by drying of the gel (third stage), and subsequently annealing treatment (fourth stage). During the annealing process, Ostwald growth at higher temperatures initiated phase transformation of the iron oxide nanoparticles and finished with the formation of the most stable phase—α-Fe_2O_3 phase. The presence of the porous, nonmagnetic matrix enables minimization of the nanoparticle interaction and enables the control of the particles size [53].

Synthesis factors of importance for every stage of the sol-gel synthesis are the choice/ratio of the precursors and pH. The influence of the variation of the synthesis conditions onto the properties of the final synthesis product is still not sufficiently investigated. What is known from the literature is that the influence of pH is reflected in the defining of the pore size. Base-catalyzed sol-gel synthesis conditioned slow hydrolysis of the alkoxide precursor and fast condensation. Final matrix pores are determined by the sizes from 2 to 50 nm [54]. In contrary, acid-catalyzed sol-gel synthesis favors rapid hydrolysis, consequently bringing to the formation of a huge number of small SiO_2 nuclei. The obtained gel consisted of the pores, with the size less than 2 nm [54]. This is explained by the influence of the hydrolysis and condensation rate on the formation of different polymers: base catalysis enabled formation of the longer, branched polymers, while acid catalysis resulted in the formation of linear polymers [55, 56]. Consequently, auto-, acid-, or base-catalyzed sol-gel syntheses could be used for the preparation of the α-Fe_2O_3/SiO_2 nanocomposite, significantly different in its properties.

5. Coercivity of the composite nanomaterials

If someone needs to get a better insight in the biomedical application of nanocomposite materials based on hematite, it is important to understand in detail synthesis, reaction mechanism, and correlation between synthesis conditions and properties of prepared samples containing nano-hematite particles. Some basic magnetic properties of nano-hematite phase are not well-established up until today, which complicate its biomedical application. From a fundamental point of view, the understanding of coercivity behavior is of great importance because coercivity presents magnetic property that significantly influences and determines application of the investigated nanomaterial.

An obstacle in a determination of the precise coercivity value of nanocomposite materials consisting hematite nanoparticles presents often occurrence of the intermediate iron oxide phase—epsilon phase (ε-Fe_2O_3) that is obtained during the synthesis of high-temperature hematite phase, by sol-gel method. ε-Fe_2O_3 polymorph is formed in the course of Fe_3O_4/γ-Fe_2O_3 $\rightarrow$ α-Fe_2O_3 structural transformation and often coexisted concomitantly with the α-Fe_2O_3 phase. The situation is additionally complicated by the inability to clearly separate a temperature range during which pure hematite or epsilon phase is formed. Pure hematite phase could be synthesized by various synthesis approaches at different temperatures up to 1100°C, while the epsilon phase is obtained only by sol-gel method and still is not prepared pure.

Having in mind that the behavior of α-Fe_2O_3 and ε-Fe_2O_3 phases is still not properly understood, the primary question regarding H_{ci} value of nanocomposite materials containing hematite nanoparticles becomes: what is the difference between the mechanism of the coercivity field variations of the nano-hematite and nano-epsilon phase, since both of these phases could be characterized by coercivity field value ranging from zero to few thousand Oe?

To answer this question, more detailed scientific research should be performed. The correlation between H_{ci} value and material microstructure is not sufficiently understood neither for bulk nor for nanomaterials [57]. Intrinsic coercivity field presents a reverse field required to reduce the magnetization (M) from the remnant magnetization (M_r) again to zero. The main problem in interpretation of the intrinsic coercivity field value is that the field measured by magnetic devices is not the coercivity field, but some critical field influenced by the magnetic interactions [57]:

$$H_{crit} = H_{ci} + H_{int} \tag{2}$$

When we deal with attempts to understand the origin of the coercivity field variations in nano-sized samples, of big importance is the independent analysis of the H_{crit} and H_{ci} values, which is difficult, since consensus about the factors that influence H_{int} and H_{ci} still is not achieved in the scientific community and presents a problem that should be overcome in the future.

The correlation between synthesis conditions and H_{ci} value occurred through the competition of different parameters, which influence and contribute in the different measure of the final H_{crit} and H_{ci} values. The mathematical expression that would describe dependence of H_{crit} on different parameters which influenced the coercivity field value has yet to be found, but roughly, it can be expressed as a function of different parameters (denoted according to the Greek alphabet) [Eq. (3)] [58]:

$$H_{crit} = f(\alpha, \beta, \gamma, \delta, \varepsilon, \eta, \theta, \iota, \kappa) \tag{3}$$

The main influence on the coercivity value arises from:

a. Synthesis conditions (parameter α)

b. Presence of different iron oxide species in the investigated sample (parameter β)

c. Contribution originated from the physical and chemical properties of the SiO_2 matrix, such as pore size distribution or the flow of different gases through the matrix (parameter γ)

d. Angular distribution of the nanoparticle orientations (parameter δ)

e. Particle size and shape distribution (parameter ε)

f. An interplay between different inter- and intra-particle interactions in the nanoparticle system (parameter η)

g. Nanoparticle structural defects (denoted as parameter θ)

h. Anisotropy field (parameter ι)

i. Surface effects (κ)

In the next section, the impact of the parameters α and β on the measured magnetic field, that is in literature labeled as coercivity field, of the synthesized nanocomposite materials will be considered. X-ray diffraction and hysteretic measurements were performed in order to investigate the influence of the variation of synthesis parameters onto the formation of hematite phase as well as on the measured magnetic field value of the samples containing the pure hematite phase and hematite phase (as a dominant phase) in combination with the epsilon phase (appeared in traces). Few examples of the peculiar H_{crit} behavior of the investigated samples will be represented.

6. α-Fe_2O_3 nanoparticles prepared by auto-, acid-, and base-catalyzed sol-gel syntheses

In the following part, synthesis of the samples prepared by auto-, acid-, and base-catalyzed sol-gel methods will be described in detail. X-ray diffraction patterns and hysteretic measurements recorded at 200 K are shown. X-ray diffraction intensity is normalized. In the entire text, the value of the magnetization (M) is normalized so that $M_s = 1$. The normalized values of magnetization were introduced in order to avoid uncertainty in the estimation of the magnetization expressed in emu/g. Having in mind that H_{crit} presents magnetic field measured by magnetometer, H_{crit} will be labeled as H_{meas} in the figures of hysteresis.

6.1 Nano-hematite-based materials prepared by auto-catalyzed sol-gel synthesis

The auto-catalyzed sol-gel synthesis implied the dissolving of iron (III) nitrate nonahydrate ($Fe(NO_3)_3 \times 9H_2O$) in water in molar ratio 0.013:1 (catalyst solution), while mixing of tetraethyl orthosilicate (TEOS), ethanol (C_2H_5OH), and water in molar proportion 1:12:12 enabled formation of alkoxide solution [59]. Solutions were mixed and stirred at room temperature. Gelation occurred during 36 days, afterward alcogel was dried for 5h at room temperature. Thermal treatment is performed in two ways. Alcogel is annealed in the air atmosphere for 3 h at 1050 and 1060°C as well as at 1050°C for 25 h. Investigated samples contained α-Fe_2O_3 as a dominant phase and smaller amount of the ε-Fe_2O_3 phase. Variation of the annealing conditions enables observation of the changes in the H_{crit} of the prepared samples, which is in detail discussed in Ref. [59].

6.1.1 Variation of the sol-gel synthesis conditions: alteration of the annealing conditions (temperature and time)

Diffraction pattern of the alcogel annealed at 1050°C for 3 h is presented in **Figure 1(a)** [59].

Hematite nanocrystallites (JCPDS card no.: 72-469) are observed as a dominant phase, while α-Fe_2O_3 phase (JCPDS card no.: 16-653) is presented in small amount. **Figure 2** presents hysteresis of the same sample [59]. Although the sample showed a higher amount of the α-Fe_2O_3 phase, measured critical field achieved a value characteristic for ε-Fe_2O_3 phase: 14.1 kOe (**Figure 2**) [59].

The increase of the annealing temperature for only 10°C (T_{ann} = 1060°C, t_{ann} = 3 h) resulted in the sharp decrease of H_{meas} (2300 Oe), which is shown in **Figure 2(a)** [59].

Observed behavior of measured critical field of the sample has been attributed to the completion of an ε-Fe_2O_3 → α-Fe_2O_3 phase transformation [59]; thus the presence of the only one phase—α-Fe_2O_3 phase—at 1060°C would be expected. With the aim to check the concluded remark regarding the completion of ε-Fe_2O_3 → α-Fe_2O_3 phase transformation at depicted temperature, investigation presented in Ref. [59] is continued by measuring the diffraction pattern of the sample annealed at 1060°C 3 h. Diffraction measurement (**Figure 2(b)**) revealed that, although the value of 2300 Oe could be characteristic for nano-sized α-Fe_2O_3 samples [24, 59], an investigated sample still contained the ε-Fe_2O_3 phase, although represented in the smaller amount than the sample annealed at 1050°C (**Figure 1(a)**).

On the other hand, published data reported more pronounced sharp decrease in H_{meas} value (600 Oe) for the sample performed to annealing at 1050°C for 25 h (**Figure 3(a)**). In order to confirm the completion of the ε-Fe_2O_3 to α-Fe_2O_3 phase transformation under these annealing conditions [59], investigation was continued by the measuring diffraction pattern of the mentioned sample (**Figure 3(b)**).

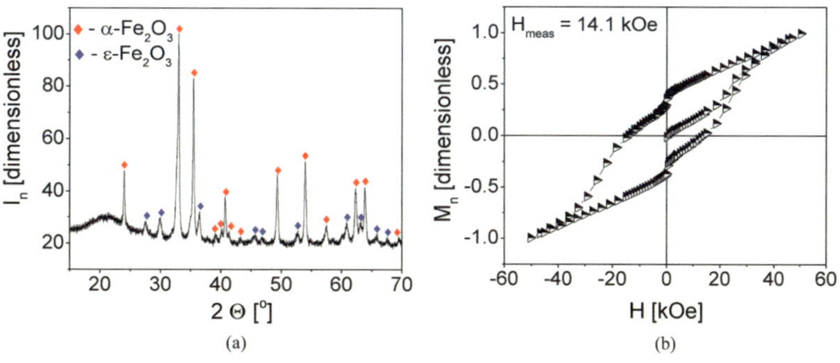

Figure 1.
Sample annealed at 1050°C for 3 h (a) diffraction pattern; (b) hysteretic curves [59].

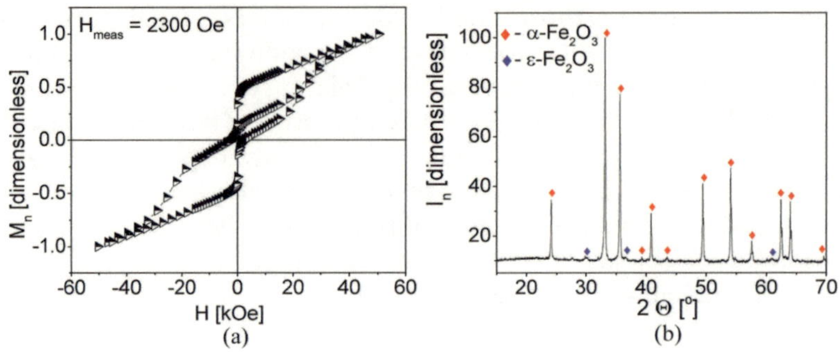

Figure 2.
Sample annealed at 1060°C for 3 h: (a) hysteretic curves [59]; (b) diffraction pattern.

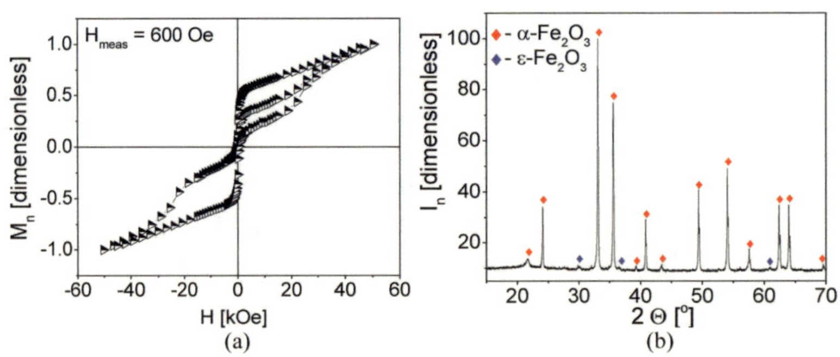

Figure 3.
Sample annealed at 1050°C for 25 h: (a) hysteretic curves [59]; (b) diffraction pattern.

In spite of a very long annealing treatment (25 h), the traces of the epsilon phase is still presented ($2\theta = 30.25°$) at the diffraction pattern presented in **Figure 3(b)**, although H_{meas} was very low for the epsilon phase, 600 Oe (**Figure 3(a)**).

Observed results pointed to the often mistakes in the scientific literature, where is sharp decrease in the measured magnetic field (H_{meas}) of the samples containing hematite and epsilon phases attributed to the finish of the ε-$Fe_2O_3 \rightarrow \alpha$-$Fe_2O_3$ phase transformation. In other words, vanishing of the huge value of the measured field (which is in the articles denoted as Hc) of the probed samples is obviously not a conclusive evidence that is pointing to the absence of epsilon phase in the sample and which could confirms completion of the ε-$Fe_2O_3 \rightarrow \alpha$-$Fe_2O_3$ transformation.

6.1.2 Variation of the sol-gel synthesis conditions: alteration of the iron precursor initial amount

Further research of the variation of the sol-gel synthesis parameters directed the investigation in the course of altering of the amount of the iron precursor, $Fe(NO_3)_3{\cdot}9H_2O$. The sample is synthesized by the identical auto-catalyzed sol-gel procedure used for the preparation of the former discussed samples, with the only difference that the molar ratio of $Fe(NO_3)_3{\cdot}9H_2O$ and water was 0.017:1 [60]. Alcogel is annealed at 1030°C for 3 h in the air atmosphere. Diffraction pattern is shown in **Figure 4(a)** [60].

Figure 4(a) revealed that α-Fe_2O_3 phase is recognized as a dominant phase, and ε-Fe_2O_3 phase as an impurity. Noteworthy, comparison of **Figures 4(a)** and **1(a)** pointed to pronounced similarities between phase compositions of the investigated samples. Surprisingly, hysteretic measurements revealed that, in spite of nearly the

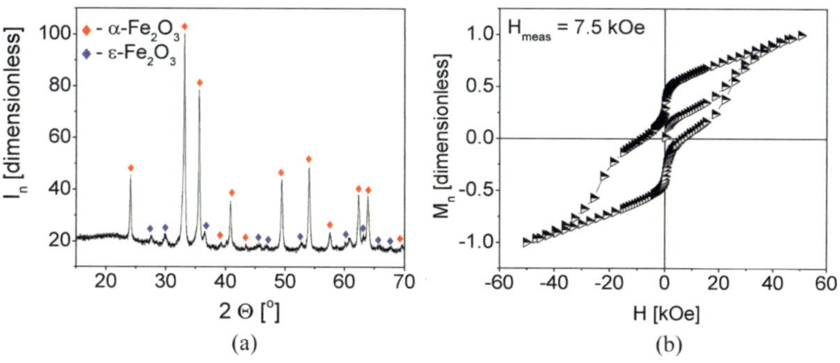

Figure 4.
Sample annealed at 1050°C for 3 h (reduced content of the iron precursor): (a) diffraction pattern; (b) hysteretic curves [59].

same diffraction patterns, the examined sample showed significantly lower value of measured magnetic field, 7.5 kOe [60]. Deeper analysis, that is out of the objective of this chapter but in detail explained elsewhere [60], showed that the origin of the H_{meas} value variations occurred as a consequence of the increased amount of the SPM nanoparticles within the investigated sample [60].

6.1.3 Variation of the sol-gel synthesis conditions: alteration of the iron (III) nitrate precursor

Hydrated iron (III) nitrate shows high non-stability and tendency to absorb humidity from the air [61]. To ensure avoidance of the contact with the air, the best way is to prepare iron (III) nitrate directly from the elemental iron and nitric acid solution. In order to investigate behavior of measured magnetic field of the sample prepared by auto-catalyzed sol-gel synthesis with the usage of anhydrous iron nitrate (instead of nonahydrated iron (III) nitrate) as a precursor, the next procedure is performed: catalyst solution is prepared by the dissolving of iron (III) nitrate in water in molar ratio 0.007:1. Alkoxide solution is prepared by mixing tetraethyl orthosilicate (TEOS), ethanol (C_2H_5OH), and water in molar proportion 1:6:6. Mixed solutions are stirred for 5 h. Gelation takes place in 20 days. Gel is dried for 19 h at 80°C and afterward is annealed at 1030°C for 3 h.

Diffraction pattern of the sample is shown in **Figure 5(a)**.

Figure 5(a) depicted the complete absence of the α-Fe_2O_3 phase and presence of the ε-Fe_2O_3 phase as the only observed iron oxide phase in the sample. Hysteretic

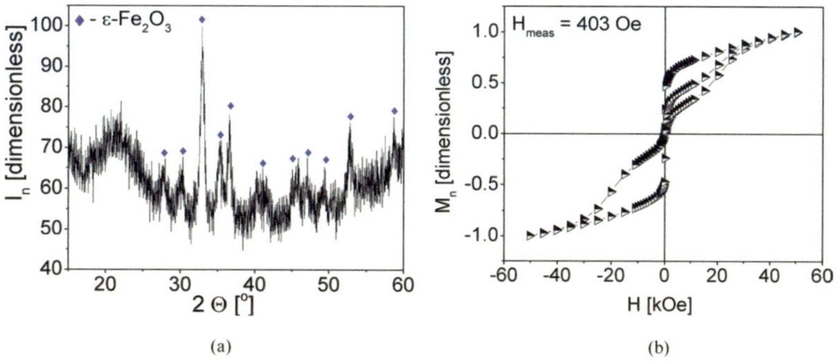

Figure 5.
Sample annealed at 1030°C for 3 h (anhydrous $Fe(NO_3)_3$ used as a precursor): (a) diffraction pattern; (b) hysteretic curves.

measurement of this sample is shown in **Figure 5(b)**. Interestingly, magnitude of the measured magnetic field of the sample was ~400 Oe. Since literature data showed that this Fe_2O_3 polymorph is characterized by high H_{ci} (10–20 kOe) [29–31] or by SPM behavior (H_{ci} ~ 0 Oe) [33], mentioned Hmeas value is not characteristic neither for high coercivity ε-Fe_2O_3 nor for the SPM ε-Fe_2O_3 phase. Moreover, obtained value is similar to the case presented in **Figure 3(b)**, where it is observed that the sample, containing the α-Fe_2O_3 as a dominant phase, showed H_{meas} ~ 600 Oe. It is important to notice here that an alcogel of the sample whose diffraction pattern is represented in the **Figure 1(a)**, performed to the thermal treatment under the similar annealing conditions, is characterized by the value of the measured magnetic field of 14.1 kOe, although hematite phase was presented as a dominant [59]. Further research of this sample will be performed by Mossbauer spectroscopy, in order to discuss the observed measured magnetic field behavior of the sample in detail.

6.2 Nano-hematite-based materials prepared by acid-catalyzed sol-gel synthesis

In order to investigate the influence of the catalyst in the sol-gel synthesis of hematite nanoparticles, a sample is synthesized by acid sol-gel synthesis route [62]. This synthesis method is similar to auto-catalyzed synthesis procedure, with the difference that nitric acid (HNO_3) is used as a catalyst. Tetraethyl orthosilicate, ethanol, iron (III) nitrate nonahydrate, and nitric acid were mixed in a molar ratio of 1:3:0.2:10. Solution is magnetically stirred for 1 h at room temperature. Gelation took place for 20 days. Obtained gel is dried at 80°C for 19 h, after which it is subjected to thermal treatment under the air atmosphere at temperature of 800°C for 2 h.

Figure 6(a) presents a diffraction pattern of the investigated sample. Pure α-Fe_2O_3 phase is observed as the only iron oxide phase. Hematite nanoparticles are observed at lower temperature than investigated examples characterized by diffraction patterns shown in **Figures 1(a)** and **4(a)**.

Corresponding hysteretic curves are shown in **Figure 6(b)**, pointing to the presence of hematite phase as the only iron oxide phase. Measured magnetic field of the α-Fe_2O_3/SiO_2 sample achieved the value of 114 Oe, which is ascribed to the coercivity of hematite nanoparticles. Accordingly, the presence of the catalyst enabling the accelerated formation of the hematite phase at lower temperatures (samples examined in **Figures 1–4** revealed the appearance of hematite phase at temperatures higher than the sample presented at **Figure 6** [59, 60]).

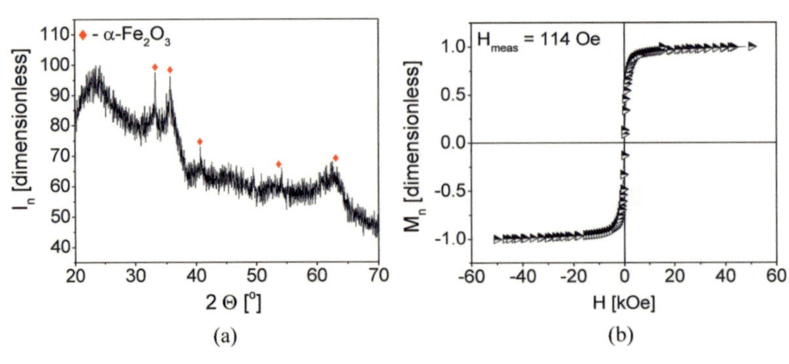

Figure 6.
Sample annealed at 800°C for 2 h (HNO_3 used as a catalyst): (a) diffraction pattern [62]; (b) hysteretic curves.

6.3 Nano-hematite-based materials prepared by base-catalyzed sol-gel synthesis

Literature review confirmed that acid-catalyzed sol-gel synthesis enabled the preparation of the samples characterized by various phase transformation routes resulting in the formation of *α-Fe_2O_3* phase: spinel phase (Fe_3O_4/γ-Fe_2O_3) → rhombohedral phase (α-Fe_2O_3) as well as spinel (Fe_3O_4/γ-Fe_2O_3) → orthorhombic (ε-Fe_2O_3) → rhombohedral (α-Fe_2O_3) phase. Dependent on the synthesis conditions, different H_{meas} values of the samples are recorded [27]. On the other hand, base-catalyzed synthesis in combination with inverse micelle method is characterized by the phase transformation route Fe_3O_4/γ-Fe_2O_3 → ε-Fe_2O_3 → α-Fe_2O_3 and presents a highly reproducible method for synthesis of high-temperature nano-hematite particles. The influence of the post-annealing treatment onto the H_{meas} value of the samples prepared by this type of sol-gel method is investigated in Ref. [63].

In the method described below, nano-hematite particles are obtained after post-annealing treatment of the samples prepared in base-catalyzed sol-gel synthesis in combination with the microemulsion method [64, 65]. Two identical microemulsions, containing water, cetyltrimethyl ammonium bromide (CTAB), buthanol, and n-octan, were mixed in a particular moral ratio. CTAB is an agent which facilitates formation of the matrix pores in the desired size [54]. Octan presents the solvent that enables the mixing of the reactants, while usage of alcohol of the somewhat longer chain (butanol) ensures a shortened time of the condensation reactions. In one microemulsion Sr^{2+} is added, whose role is the acceleration of the particle growth along one crystallographic axis, resulting in the rod morphology of the nanoparticles [64]. In another microemulsion a base catalyst, ammonia, is added that possesses a significant role in the defining of the SiO_2 pore size. Mixing the microemulsions enables stirring of the solution. Afterwards, TEOS is added in the precise stechiometrical ratio. Gao et al. [66] confirmed that the ideal volume ratio of the TEOS and alcohol (desirable in order to shorten the gelation time) is 1:2, while the same effect is achieved by simultaneously mixing the TEOS and NH_3 in the volume ratio 2:5.

6.3.1 Variation of the sol-gel synthesis conditions: post-annealing treatment

Having in mind that the influence of the annealing conditions on the samples synthesized by this method is well-established in literature [64, 65], obtained samples were performed to post-annealing treatment in order to investigate coercivity behavior of the samples post-annealed at low temperature (100°C) and high temperature (1100°C) [63, 67].

The synthesis of the sample comprised the preparation of two identical micelles, consisting of CTAB, isooctane, butanol, and water in the molar ratio—0.03:0.33:0.12:1.00. Iron (III) nitrate (prepared by dissolving elemental iron in nitric acid and water) is added to the water in the molar ratio 0.00047:1. In the first micelles precursors of the iron and strontium ions in molar ratio 3:1 are added. In another micelle 0.09 mol of ammonia is added. After mixing the micelles, TEOS is dropped into the stirred solution (volume ratio of TEOS and ammonia was 4.5:1.7, while volume ratio of the TEOS and butanol was 4.5:10.7). Solution is stirred for 24 h. Afterward, precipitate is collected and treated with a chloroform and ethanol in order to wash organic moistures, attached to the surface of the precipitated nanoparticles. A coprecipitate has been annealed at 1050°C for 4 h. The same amounts of the sample are performed to the post-annealing treatment [63].

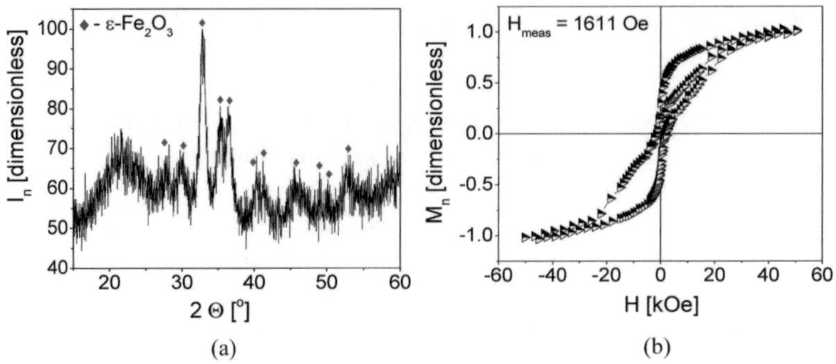

Figure 7.
Sample annealed at 1050°C for 4 h, post-annealed at 100°C for 3 h: (a) diffraction pattern; (b) hysteretic curves [60].

Diffraction pattern of the sample post-annealed at 100°C is shown in **Figure 7(a)** [67]:

The only noticed phase was the ε-Fe_2O_3 phase. In order to examine its magnetic behavior, hysteretic curves are measured and shown in the **Figure 7(b)** [67]. Measured magnetic field of the sample was 1611 Oe. Literature data showed that obtained value is not exactly characteristic for epsilon phase coercivity, and it would be more appropriate to ascribe that H_{meas} value to hematite phase, then for epsilon. According to **Figure 7(b)** and the literature data [67], it can be concluded that post-annealing treatment at 100°C brings to the drastic drop of measured measured magnetic field value. Before post-annealing treatment, sample annealed at 1050°C for 4 h showed measured magnetic field value of 21.3 kOe [63], although the phase composition of the sample was the same [67]. This fact underlined that sharp changes in the H_{meas} value of the nanocomposite samples prepared by sol-gel method could not be ascribed to Fe_2O_3 polymorph transformations.

To get a better insight in the H_{meas} variations initiated by post-annealing treatment, the piece of alcogel annealed at 1050°C for 4 h was performed to the post-annealing treatment at 1100°C for 3 h [63].

Figure 8(a) revealed that the only observed iron oxide phase is pure α-Fe_2O_3 phase [63], pointing to the ending of the ε-Fe_2O_3 to α-Fe_2O_3 phase transformation. In diffraction pattern is in addition noticed change of the silica matrix, converted from amorphous silica to highly crystalline cristoballite (JCPDS card no.: 39-1425) and quartz phase (JCPDS card no.: 46-1045) [56, 63]. Interestingly, hysteretic curves shown in **Figure 8(b)** exerted measured magnetic field value of 1760 Oe

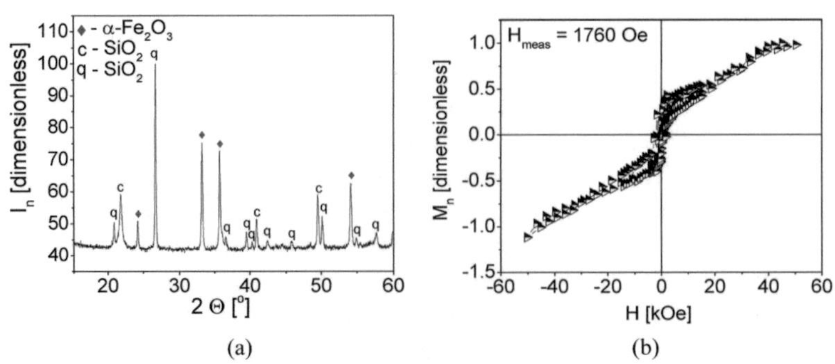

Figure 8.
Sample annealed at 1050°C for 4 h, post-annealed at 1100°C for 3 h: (a) diffraction pattern (letter "c" referred to cristoballite SiO_2 phase; letter "q" referred to quartz SiO_2 phase); (b) hysteretic curves [49].

that is quite similar to the measured value of the sample post-annealed at 100°C (H_{meas} ~ 1600 Oe). Noteworthy, **Figures 7(b)** and **8(b)** showed a similar values of measured magnetic field of the samples consisted of obviously different phases (**Figures 7(a)** and **8(a)**), pointing to the potential usage of the samples consisted of α-Fe_2O_3 and ε-Fe_2O_3 together (as in the case of "SPION" species), although this claim requires deeper investigation of magnetic properties of the samples.

Conclusively, it is important to notice that all hysteretic curves (except **Figure 6(b)**), referring to the sample containing the only hematite phase, observed at lower temperature in comparison with the other investigated samples) could be classified as hysteretic loops having constricted middles (wasp-waisted loops). Generation of wasp-waisting curves appears as a result of two population of the particles characterized by distinct coercivity spectra; numerical simulations reveals that wasp-waisting curves requires and SPM contribution [68], that is confirmed by **Figures 4(b)** and **1(b)** [58]. Experimental results shown in **Figures 2(a)** and **3(a)** revealed that the decrease of measured magnetic field value of the samples is not a certain parameter that indicates vanishing of the other iron oxide polymorphs and the presence of the pure hematite nanoparticles. Moreover, represented results pointed out that ε-Fe_2O_3 → α-Fe_2O_3 phase transformation cannot be the decisive factor on the coercivity value of the nanocomposite material (**Figures 1–4**) [58]. The variation of the initial iron ion precursor amount [60] enables the alteration of measured magnetic field value without changing the phase composition of the nanocomposite material. The usage of anhydrous Fe^{3+} precursor and change of the precursor ratio during synthesis cause the significant alterations of the phase composition of the investigated samples and differences in H_{meas} value (**Figures 1** and **5**). Appearance of the catalyst accelerated phase transformations of the Fe_2O_3 polymorph and favors the obtaining of the pure nano-hematite phase (**Figure 6(a)**).

Noteworthy, nanocomposite samples containing different phase compositions could be characterized by a significantly similar measured magnetic field values. This is confirmed by examination of the samples prepared by auto-catalyzed sol-gel synthesis (**Figures 3** and **5**) and samples synthesized by base-catalyzed sol-gel method (**Figures 7** and **8**). An observed feature is pointing to the fact that coercivity of the nanocomposite materials could not be mainly driven by the parameters α and β from Eq. (2). The results represented in this chapter indicated the necessity of taking other parameters [parameters: γ − κ, Eq. (2)] into consideration, in order to understand properly the coercivity behavior of nanocomposite materials.

6.4 TEM measurements

An important issue in the characterization of the nanoparticles presents transmission electron microscopy (TEM). In order to understand the distribution of

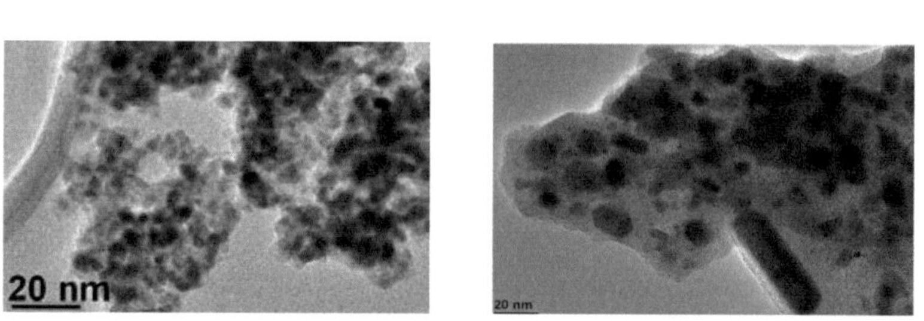

(a) (b)

Figure 9.
TEM micrographs of the sample: (a) prepared by auto-catalyzed sol-gel synthesis, annealed at 1050°C for 3 h [52], left; (b) prepared by base-catalyzed sol-gel synthesis, annealed at 1050°C for 4 h [56], right.

nano-hematite particles, TEM micrographs of the chosen nanocomposite samples annealed at 1050°C, obtained by auto-catalyzed and base-catalyzed sol-gel synthesis routes, are shown in **Figure 9(a)** and **(b)**.

Detailed TEM analysis is given in Ref. [59, 63]. Quantitative description of morphological properties of the investigated particles is performed by measuring ellipticity. The results of the analysis showed that the shape of observed Fe_2O_3 nanoparticles (ε-Fe_2O_3 and α-Fe_2O_3) varies from ellipticity to circularity. In other words, **Figure 9** confirms the presence of nonideally spherical particles, whose shape deviates from circularity in a different measure [59]. Fe_2O_3 particle sizes, presented in **Figure 9(a)**, are ranging between 10 and 20 nm, while the sample presented in **Figure 9(b)** showed a wider particle size distribution, from 4 to 50 nm, and the same variations from ellipticity to circularity. Wide size distribution leads to the presence of different particle shapes during the annealing treatment, elliptic/spherical (**Figure 9(a)**). This feature appeared as a consequence of the fact that sol-gel method consisted of coprecipitation of the particles within the SiO_2 pores (coprecipitated samples are characterized by wide particle size distribution) [53]. The best way to overcome the mentioned problem is the coating of the nanoparticles within the SiO_2 pores [69].

Notwithstanding, a significant difference between micrographs is the presence of nanorod particles within the sample synthesized by base-catalyzed sol-gel synthesis (**Figure 9(b)**). Rod-like morphology appeared as a consequence of the participation of group II element, Sr^{2+}, in the synthesis procedure (**Figure 9(b)**). The addition of Sr^{2+} ions accelerated the growth of the ε-Fe_2O_3 particles in one crystallographic axis [63], inducing more pronounced shape variations and formation of rod-like nanoparticles. If we recall the fact that TEM image shown in **Figure 9(b)** revealed the presence of the only γ-Fe_2O_3 and ε-Fe_2O_3 nanoparticles in the investigated sample (discussed in more detail in the Ref. [63]), as well as having in mind that α-Fe_2O_3 formation occurs as a consequence of phase transformations ε-$Fe_2O_3 \rightarrow \alpha$-$Fe_2O_3$, it can be assumed that the sample presented in **Figure 8** contained rod-like α-Fe_2O_3 nanoparticles.

It is important to note that the origin of dependence of H_{meas} behavior on the synthesis conditions of the samples investigated in this chapter is found in quantum mechanics.

Briefly, the quantity that strongly affects the shape of hysteresis loop is magnetic anisotropy [parameter ι, in Eq. (3)]. For the most simplest case, in crystal systems whose symmetry is determined by a single axis of high symmetry (uniaxial symmetry), anisotropy energy is defined as:

$$E_a \sim KV\sin^2\theta \qquad (4)$$

where K is the anisotropy constant, V is the volume, and θ is the angle between two spins with respect to each other [17]. The overall magnetic anisotropy energy is dependent on the symmetry of the investigated systems and defined by various contributions, such as magnetocrystalline anisotropy, shape anisotropy, surface anisotropy, strain anisotropy and stress anisotropy.

Anisotropy energy appeared as a consequence of spin-orbit interaction and the partial quenching of the angular momentum [17]. From the aspect of nanomaterial preparation and dependence of samples of magnetic properties on synthesis conditions (annealing temperature and time), it is important to emphasize that the anisotropy constant is strongly temperature dependent [17]. Independent of the presence of the same or different iron oxide polymorph phases within the sample, differences in the structure and morphology characteristics of each individual nanoparticle resulted in the changes in a magnetic anisotropy.

Noteworthy, alteration of the SiO_2 matrix during the annealing treatment impacts magnetic properties of the samples [59]. Gas diffusion in the SiO_2 matrix, initiated by thermal treatment, caused the crystallization of the amorphous matrix and changes in the size of the pores. Coalescing of the pores resulted in the alterations of the distance between magnetic nanoparticles within the pores [59], consequently influencing the magnetic behavior of the samples.

7. Conclusion

The main message of this chapter was to emphasize the importance of the investigation of the influence of the synthesis parameter variations onto the magnetic properties of the composite materials containing nano-hematite particles that could be used as a starting material for preparation of multifunctional nanoparticles, used in different areas of biomedicine. Since coercivity field presents a parameter of importance for application of this type of materials, alterations of measured measured H_{crit} value, initiated by changing the synthesis parameters, are discussed. To get a better insight into relation between synthesis conditions and magnetic properties of composites containing α-Fe_2O_3 nanoparticles, sol-gel synthesis is recognized as a suitable preparation method. Alterations of measured H_{crit} value of the samples are driven by the variation of the pH of the performed sol-gel synthesis (auto-, acid-, or base-catalyzed), initial Fe^{3+} and Si^{4+} precursor ratio, amount of the iron precursor, and annealing conditions (T and t) and by performing post-annealing treatment. The author expected that this chapter will facilitate a current and objective evaluation of the knowledge regarding the search for the exact mathematical expression of the measured intrinsic coercivity field value of the composite nanomaterials containing nano-hematite phase, which is of significance for improvement of the preparation of a high-quality nano-α-Fe_2O_3 particles for biomedical application.

Acknowledgements

The research was carried out thanks to the support of the Ministry of Education, Science and Technology Development, Republic of Serbia (Project No. III 45015). The author gratefully acknowledges Dr. Vojislav Spasojević for the magnetic measurement (**Figure 5(a)**).

Author details

Violeta N. Nikolić
Laboratory for Theoretical Physics and Condensed Matter Physics, Institute of Nuclear Sciences Vinča, University of Belgrade, Serbia

*Address all correspondence to: violeta@vinca.rs

References

[1] Cornell RM, Schwertmann U. The Iron Oxides: Structure, Properties, Reactions, Occurrence and Uses. Weinheim, Germany: VCH; 2003

[2] Glasby GP, Schulz HD. Eh-Ph diagrams for Mn, Fe, Co, Ni, Cu and as under seawater conditions: Application of two new types of eh ph diagrams to the study of specific problems in marine geochemistry. Aquatic Geochemistry. 1999;**5**(2):227-248

[3] Ling Y, Wheeler DA, Zhang JZ, Li Y. Optical properties and applications of hematite (α-Fe_2O^3) nanostructures. In: One-Dimensional Nanostructures: Principles and Applications. New York: John Wiley & Sons, Inc; 2012. pp. 167-184

[4] Kefeni KK, Msagati TA, Nkambule TT, Mamba BB. Synthesis and application of hematite nanoparticles for acid mine drainage treatment. Journal of Environmental Chemical Engineering. 2018;**6**(2):1865-1874

[5] Shinde SS, Bansode RA, Bhosale CH, Rajpure KY. Physical properties of hematite α-Fe_2O_3 thin films: Application to photoelectrochemical solar cells. Journal of Semiconductors. 2011;**32**(1):013001-013003

[6] Morel D. Hematite: Sources, Properties and Applications. New York: Nova Science Publishers Incorporated; 2013

[7] Ling Y, Li Y. Review of Sn-doped hematite nanostructures for photoelectrochemical water splitting. Particle & Particle Systems Characterization. 2014;**31**(11):1113-1121

[8] Rao BN, Padmaraj O, Kumar PR, Venkateswarlu M, Rao VM, Satyanarayana N. Synthesis of hematite α-Fe_2O_3 nanospheres for lithium ion battery applications. AIP Conference Proceedings. 2015;**1665**(1):060014

[9] Kurata T, Katamoto T, Hiriishi N, Kiyama M. U.S. Patent No. 5,133,805. Washington, DC: U.S. Patent and Trademark Office; 1992

[10] Oliveira HS, Almeida LD, de Freitas VA, Moura FC, Souza PP, P P, et al. Nb-doped hematite: Highly active catalyst for the oxidation of organic dyes in water. Catalysis Today. 2015;**240**(1):176-181

[11] Pokropivny VV, Skorokhod VV. Classification of nanostructures by dimensionality and concept of surface forms engineering in nanomaterial science. Materials Science and Engineering: C. 2007;**27**(5-8):990-993

[12] Bodker F, Hansen MF, Koch CB, Lefmann K, Morup S. Magnetic properties of hematite nanoparticles. Physical Review B. 2000;**61**(10):6826-6829

[13] Catti M, Valerio G, Dovesi R. Theoretical study of electronic, magnetic and structural properties of α-Fe_2O_3 (hematite). Physical Review B. 1995;**51**(12):6789-6921

[14] Zhu M, Wang Y, Meng D, Qin X, Diao G. Hydrothermal synthesis of hematite nanoparticles and their electrochemical properties. Journal of Physical Chemistry C. 2012;**116**(30):16276-16285

[15] Li X, Zhao T, Sun L, Aifantis K, Fan Y, Feng Q, et al. The applications of conductive nanomaterials in the biomedical field. Journal of Biomedical Materials Research Part A. 2016;**104**(1):322-339

[16] Oles A, Kajzar F, Jucab M, Sikora W. Magnetic Structures

Determined by Neutron Diffraction. Warszava-Krakov: Panstwore Wydawnicto Naukome; 1976

[17] Blundell S. Magnetism in Condensed Matter. New York: Oxford University Press, Inc; 2003

[18] Kittel C. Introduction to Solid State Physics. New York: John Willey & Sons; 1996

[19] Klabunde NA. Nanoscale Materials in Chemistry. New York: John Willey & Sons; 2010

[20] Xu S, Habib AH, Gee SH, Hong YK, McHenry ME. Spin orientation, structure, morphology, and magnetic properties of hematite nanoparticles. Journal of Applied Physics. 2015;**117**(17):17A315. DOI: 10.1063/1.4914059

[21] Klein S, Sommer A, Distel LV, Neuhuber W, Kryschi C. Superparamagnetic iron oxide nanoparticles as radiosensitizer via enhanced reactive oxygen species formation. Biochemical and Biophysical Research Communications. 2012;**425**(2):393-397

[22] Barhoumi L, Dewez D. Toxicity of superparamagnetic iron oxide nanoparticles on green alga Chlorella vulgaris. BioMed Research International. 2013;**2013**:1-11. DOI: 10.1155/2013/647974

[23] Hill AH, Jiao F, Bruce PG, Harrison A, Kockelmann W, Ritter C. Neutron diffraction study of mesoporous and bulk hematite α-Fe_2O_3. Chemistry of Materials. 2008;**20**(1):4891-4899

[24] Tadic M, Citakovic N, Panjan M, Stanojevic B, Markovic D, Jovanovic D, et al. Synthesis, morphology and microstructure of pomegranate-like hematite (α-Fe_2O_3) superstructure with high coercivity. Journal of Alloys and Compounds. 2012;**543**(5):118-124

[25] King JW, Channel ET. Sedimentary magnetism, environmental magnetism, and magnetostratigraphy. International Union of Geodesy and Geophysics. 1987;**29**(1):358-360

[26] Kurmoo M, Rehspringer JL, Hutlova A, Dorleans C, Vilminot S, Estournes C, et al. Formation of nanoparticles of ε-Fe_2O_3 from yttrium iron garnet in a silica matrix: Unusually hard magnet with a Morin-like transition below 150 K. Chemistry of Materials. 2005;**17**(1):1106-1114

[27] Popovici M, Gich M, Niznansky D, Roig A, Savii C, Casas L, et al. Optimized synthesis of the elusive ε-Fe_2O_3 phase via sol-gel chemistry. Chemistry of Materials. 2004;**25**(1):5542-5548

[28] Gich M, Gazquez J, Roig A, Crespi A, Fontcuberta J, Idrobo JC, et al. Epitaxial stabilization of ε-Fe_2O_3(00l) thin films on $SrTiO_3$(111). Applied Physics Letters. 2010;**96**(1):112508-112511

[29] Barick KC, Varaprasad BCS, Bahadur D. Structural and magnetic properties of γ- and ε-Fe_2O_3 nanoparticles dispersed in silica matrix. Journal of Non-Crystalline Solids. 2010;**356**(1):153-159

[30] Yakushkin SS, Balaev DA, Dubrovskiy AA, Semenov SV, Shaikhutdinov KA, Kazakova MA, et al. Evolution of the Fe^{3+} ion local environment during the phase transition ε-$Fe_2O_3 \rightarrow \alpha$-$Fe_2O_3$. Journal of Superconductivity and Novel Magnetism. 2018;**31**(4):1209-1217

[31] Liu J, Qiao SZ, Chen JS, Lou XWD, Xing X, Lu GQM. Yolk/shell nanoparticles: New platforms for

nanoreactors, drug delivery and lithium-ion batteries. Chemical Communications. 2011;**47**(47):12578-12591

[32] Lu C, Liu X, Li Y, Yu F, Tang L, Hu Y, et al. Multifunctional janus hematite-silica nanoparticles: Mimicking peroxidase-like activity and sensitive colorimetric detection of glucose. ACS Applied Materials & Interfaces. 2015;7(28):15395-15402

[33] Yu BY, Kwak SY. Carbon quantum dots embedded with mesoporous hematite nanospheres as efficient visible light-active photocatalysts. Journal of Materials Chemistry. 2012;**22**(17):8345-8353

[34] Yin ZF, Wu L, Yang HG, Su YH. Recent progress in biomedical applications of titanium dioxide. Physical Chemistry Chemical Physics. 2013;**15**(14):4844-4858

[35] Mirzaei A, Janghorban K, Hashemi B, Bonyani M, Leonardi SG, Neri G. Highly stable and selective ethanol sensor based on α-Fe_2O_3 nanoparticles prepared by Pechini sol–gel method. Ceramics International. 2016;**42**(5):6136-6144

[36] Selvam AP, Muthukumar S, Kamakoti V, Prasad S. A wearable biochemical sensor for monitoring alcohol consumption lifestyle through ethyl glucuronide (EtG) detection in human sweat. Scientific Reports. 2016;**6**(2):23111-23114

[37] Kim DH, Shim YS, Jeon JM, Jeong HY, Park SS, Kim YW, et al. Vertically ordered hematite nanotube array as an ultrasensitive and rapid response acetone sensor. ACS Applied Materials & Interfaces. 2014;**6**(17):14779-14784

[38] Rohr TE. U.S. Patent No. 5,445,971. Washington, DC: U.S. Patent and Trademark Office; 1995

[39] Litvinov D, Willson R. U.S. Patent No. 8,456,157. Washington, DC: U.S. Patent and Trademark Office; 2013

[40] Bhar R, Kaur G, Mehta SK. Exploring drying pattern of a sessile droplet of genomic DNA in the presence of hematite nanoparticles. Scientific Reports. 2018;**8**(1):6352-6355

[41] Lu Y, Yin Y, Mayers BT, Xia Y. Modifying the surface properties of superparamagnetic iron oxide nanoparticles through a sol–gel approach. Nano Letters. 2002;**2**(3):183-186

[42] Cui H, Liu Y, Ren W. Structure switch between α-Fe_2O_3, γ-Fe_2O_3 and Fe_3O_4 during the large scale and low temperature sol–gel synthesis of nearly monodispersed iron oxide nanoparticles. Advanced Powder Technology. 2013;**24**(1):93-97

[43] Gupta AK, Gupta M. Synthesis and surface engineering of iron oxide nanoparticles for biomedical applications. Biomaterials. 2005;**26**(18):3995-4021

[44] Li Z, Barnes JC, Bosoy A, Stoddart JF, Zink JI. Mesoporous silica nanoparticles in biomedical applications. Chemical Society Reviews. 2012;**41**(7):2590-2605

[45] Tartaj P, del Puerto Morales M, Veintemillas-Verdaguer S, Gonzalez-Carreno T, Serna CJ. The preparation of magnetic nanoparticles for applications in biomedicine. Journal of Physics D: Applied Physics. 2003;**36**(13):R182-R197

[46] Salata OV. Applications of nanoparticles in biology and medicine. Journal of Nanobiotechnology. 2004;**2**(1):3-6

[47] Jesus M, Penades S. Glyconanoparticles: Types, synthesis and applications in glycoscience, biomedicine and material science.

Biochimica et Biophysica Acta (BBA)-General Subjects. 2006;**1760**(4):636-651

[48] Neuberger T, Schopf B, Hofmann H, Hofmann M, Von Rechenberg B. Superparamagnetic nanoparticles for biomedical applications: Possibilities and limitations of a new drug delivery system. Journal of Magnetism and Magnetic Materials. 2005;**293**(1):483-496

[49] Nitin N, LaConte LEW, Zurkiya O, Hu X, Bao G. Functionalization and peptide-based delivery of magnetic nanoparticles as an intracellular MRI contrast agent. Journal of Biological Inorganic Chemistry. 2004;**9**(6):706-712

[50] Lopez-Sanchez J, Serrano A, Del Campo A, Abuin M, Rodriguez de la Fuente O, Carmona N. Sol–gel synthesis and micro-raman characterization of ε-Fe_2O_3 micro-and nanoparticles. Chemistry of Materials. 2016;**28**(2):511-518

[51] Taboada E, Gich M, Roig A. Nanospheres of silica with an ε-Fe_2O_3 single crystal nucleus. ACS Nano. 2009;**3**(11):3377-3382

[52] Akbar S, Hasanain SK, Azmat N, Nadeem M. Synthesis of Fe_2O_3 nanoparticles by new sol-gel method and their structural and magnetic characterizations. arXiv preprint cd-mat/04084802004. pp. 1-9

[53] Zhang J, Chen X, Shen Y, Li Y, Hu Z, Chu J. Synthesis, surface morphology, and photoluminescence properties of anatase iron-doped titanium dioxide nano-crystalline films. Physical Chemistry Chemical Physics. 2011;**13**(28):13096-13105

[54] Stober W, Fink A, Bohn E. Controlled growth of monodisperse silica spheres in the micron size range. Journal of Colloid and Interface Science. 1968;**26**(1):62-69

[55] Brinker CJ, Scherer GW. Sol-Gel Science, The Physics and Chemistry of Sol-Gel Processing. New York: Academic Press; 1990

[56] Nikolic VN. Магнетне особине наночестица оксида гвожђа површински модификованих силицијум диоксидом и олеинском киселином [PhD dissertation]. Serbia: INN "Vinca", University of Belgrade; 2017

[57] Givord D, Rossignol M, Taylor D. Coercivity mechanisms in hard magnetic materials. Journal de Physique IV Colloque. 1992;**02**(C3):95-98

[58] Nikolic VN, Milic MM, Zdravkovic JD, Spasojevic V. Origin of the intrinsic coercivity field variations and magnetic study of ε-Fe_2O_3 phase. Russian Journal of Physical Chemistry A. 2019;**93**(3):588-593

[59] Nikolic VN, Tadic M, Panjan M, Kopanja L, Cvjeticanin N, Spasojevic V. Influence of annealing treatment on magnetic properties of Fe_2O_3/SiO_2 and formation of ε-Fe_2O_3 phase. Ceramics International. 2016;**43**(2):3147-3155

[60] Nikolic VN, Milic MM, Zdravkovic JD, Spasojevic V. Origin of the intrinsic coercivity field variations of ε-Fe_2O_3. Russian Journal of Physical Chemistry A. 2019;**93**(2):377-383

[61] Schmidt H, Asztalos A, Bok F, Voigt W. New iron (III) nitrate hydrates: Fe $(NO_3)_3 \cdot xH_2O$ with x= 4, 5 and 6. Acta Crystallographica Section C: Crystal Structure Communications. 2012;**68**(6):29-33

[62] Nikolic VN, Spasojevic V. Analysis of ZFC/FC curves of hematite-silica nanocomposite materials. In: 8th. Belgrade, Serbia: International Conference on Defensive Technologies; 2018

[63] Nikolic VN, Spasojevic V, Panjan M, Kopanja L, Mrakovic A, Tadic M. Re-formation of metastable

ε-Fe_2O_3 in post-annealing of Fe_2O_3/SiO_2 nanostructure: Synthesis, computational particle shape analysis in micrographs and magnetic properties. Ceramics International. 2017;**43**(3):7497-7507

[64] Sakurai S, Tomita K, Hashimoto K, Yashiro H, Ohkoshi S. Preparation of the nanowire form of ε-Fe_2O_3 single crystal and a study of the formation process. Journal of Physical Chemistry C. 2008;**112**(5):20212-20216

[65] Sakurai S, Jin J, Hashimoto K, Okhoshi S. Reorientation phenomenon in a magnetic phase of ε-Fe_2O_3 Nanocrystal. Journal of the Physical Society of Japan. 2005;**74**(7):1946-1949

[66] Gao H, Yang J. Nanoscale silicon dioxide prepared by sol-gel process. Modern Applied Science. 2010;**4**(9):152-156

[67] Nikolic VN, Tadic M, Mrakovic A, Spasojevic V. Unusual temperature dependence of coercivity in ε-Fe_2O_3 phase. In: 9th International Conference on Nanomaterials-Research & Application; Brno, Czech. 2017

[68] Tauxe L, Mullender TAT, Pick T. Potbellies, wasp-waists, and superparamagnetism in magnetic hysteresis. Journal of Geophysical Research: Solid Earth. 1996;**101**(B1):571-583

[69] Guo XH, Deng YH, Gu D, Che RC, Zhao DY. Synthesis and microwave absorption of uniform hematite nanoparticles and their core-shell mesoporous silica nanocomposites. Journal of Materials Chemistry. 2009;**19**(2):6706-6712

Chapter 9

Future of Nanoparticles in the Field of Medicine

Neha Sharma

Abstract

The chapter deals with the application of iron oxide nanoparticles in the field of medicine. It focuses on the treatment of cancerous cells in the body as a case study. Cancer as we all know is a disease which is spreading at the speed of light across the nations, primarily due to the lifestyles and heredity. The human war against the disease is on, and many cures are in practice or under research, so as to limit the deaths due to it. Most of the research is focused on finding alternative and effective techniques in conquering cancer, so that the stigma attached with it can be diminished; the researchers are also focusing on lowering the side effects of the currently practiced cures. We all hope that a day will come when it will come under the category of conquerable diseases. It has been shown that cancer deaths in the world have declined considerably, but it is still unconquerable. It is still one of the leading causes of death around the globe. Usual therapy like radiation, surgery, and immunotherapy in addition to chemotherapy has shown challenges like ease of access to the tumor cells, danger of operating on a vital organ to name some. Off late, research laboratories are using nanoparticles for the detection in addition to drug delivery in treatment of various diseases. It gives boost to minimizing the side effects encountered in conventional therapies at the cellular and tissue level. Nanoparticles' widespread use is accounted by their size.

Keywords: magnetic nanoparticles, tumor, hyperthermia

1. Introduction

The discussion in the succeeding paragraphs is not limited to only iron oxide nanoparticles, it deals with all nanoparticles which are magnetic in nature; hence, there is a reoccurrence of the phrase "magnetic nanoparticles."

The technique or the procedure which I am going to discuss is widely known as hyperthermia. It is the phenomenon which involves selective heating of magnetic particles using high-frequency magnetic field. The case presented here uses the fact that the tumor in the affected area can be removed by heating it up to certain temperatures depending on the different parameters of the nanoparticles. The whole idea started with the introduction of the magic bullet, a concept given by Nobel laureate Paul Ehrlich (1854–1915), 1908, in the field of medicine in the field of immunity. The idea of Magic Bullet projected by selective targeting of disease causing organism in addition to delivery of toxin for the affected area. The procedure suggested was to first identify the cancerous cell/tissue and then target the magic bullet of nanoparticles at the site and the blast the cell for destroying the un-repairable cell or delivering dose with the help of nanoparticles' magic bullet, which at the site will open to deliver the drug.

Later, after the invention of the magic bullet, a number of hyperthermia techniques were suggested since 1970. Scientists Zimmermann and Pilwat in 1976 proposed the use of magnetic erythrocytes for delivery of drug at the affected site. Research group led by Freeman et al. in 1960 came up with the idea wherein magnetic nanoparticles could be transported through the vascular system and grouped in a specific part of the body using magnetic field. As recently as 2009, Boris Polyak and Gary Friedman explored clinical potential and applications of magnetic targeting for site-specific drug delivery.

The treatment of hyperthermia involves heating of injected cancer-specific biomolecules coated with magnetic nanoparticles at the affected area. It involves selective heating of magnetic particles, which are positioned at the affected site, using high-frequency magnetic field. Removal of tumor (different diameter sizes) located in the liver is studied by varying power applied for different exposure times theoretically using heating model given by Tsafnat et al.

I again present the research by my group wherein we optimized the power requirements for the destruction of diseased cell at the location. The entire work was around the concept of the treatment of tumor using hyperthermia, which involves heating of injected cancer-specific biomolecules coated with magnetic nanoparticles at the affected area. The procedure involved selective heating of magnetic nanoparticles, which are positioned at the affected site of the diseased cell, using a very high-frequency magnetic field. The hypothesis used is that the tumor in the affected area can be removed by heating it up to temperatures, in the range of 41–46°C based on earlier research in the area. It was proposed that a tumor with a diameter size of 5 cm can be efficiently removed, if magnetic nanoparticles (present at the tumor site) were exposed for 10 s, with a power range of 2.75–6.5 W.

2. Approach

Based on the different research models given, we used the theory according to which the tumor in the affected area can be removed by heating it up to temperatures [1–20], in the range of 41–46°C. It was proposed that if the power ranges 2.75–6.5 W, when applied to the magnetic nanoparticles (present at the tumor site), for duration of (up to) 10 s, a tumor having diameter of size 5 cm can be successfully/efficiently removed. Dependency of temperature in the affected area on the diameter of the magnetic nanoparticle in addition to exposure time of magnetic nanoparticles by alternating magnetic field and power was studied [20, 21].

The work tries to carry forward the results presented in orphan drugs [22], which proved that, if heat is applied for a duration of 10 s for an applied power to magnetic nanoparticles of 6.5 W, it leads to removal of tumor (up to radius size of 10 cm). We explore the variation in applied power within which the desired results can be obtained, which in turn leads to minimizing the running cost and undue heating of healthy tissue/cell in the vicinity of the affected area. The study presented is based on the model suggested by Tsafnat et al. [19] for heating of liver tumors. According to their study, the affected area demonstrates lower levels of blood perfusion than a healthy one. This results in partial safety for the healthy liver tissue during localized hyperthermia treatment. The model (mathematical) reproduced here simulates the practical heat diffusion from the affected area (tumor) to its surrounding (unaffected) tissue. It is assumed that temperatures in the two respective areas have an effect on each other at the boundary interface.

3. The mathematical model

It is assumed that the shape of the tumor is a spherical tissue with radius a which is surrounded by a normal tissue again assumed to be a bigger concentric sphere with radius b (see **Figure 1**).

The model presented also works on a hypothesis that the heat source with constant power density P is concentrated within the small sphere of radius a surrounded by a medium of homogenous heat conductivity.

As the model under study is having the spherical symmetry and the homogenous time-independent power density P inside the sphere, the temperature distribution is a function of distance r from the center of the sphere and on time t.

The differential equations of heat conduction [11, 22, 23] are used for defining the required mathematical model given by [11, 22, 23]

$$c_1\rho_1\frac{\partial T_1}{\partial t} = \frac{k_1}{r^2}\frac{\partial}{\partial r}\left(r^2\frac{\partial T_1}{\partial r}\right) + P, \text{ for } 0 \le r < a, \text{ interior of tumor} \tag{1}$$

$$c_2\rho_2\frac{\partial T_2}{\partial t} = \frac{k_2}{r^2}\frac{\partial}{\partial r}\left(r^2\frac{\partial T_2}{\partial r}\right) \text{ for } a \le r \le b, \text{ exterior of tumor} \tag{2}$$

The subscript "1" is used to represent tumor tissue, while the subscript "2" refers to normal tissue, and the various parameters in these equations are defined as follows:

T is for the temperature.

c represents specific heat capacity.

ρ represents the density.

k is the heat conductivity.

Based on the hypothesis that the temperature and flux at the boundary are continuous and fine, the boundary conditions can be written as

$$T_1(a,t) = T_2(a,t) \tag{3}$$

$$k_1\frac{\partial T_1(a,t)}{\partial r} = k_2\frac{\partial T_2(a,t)}{\partial r} \tag{4}$$

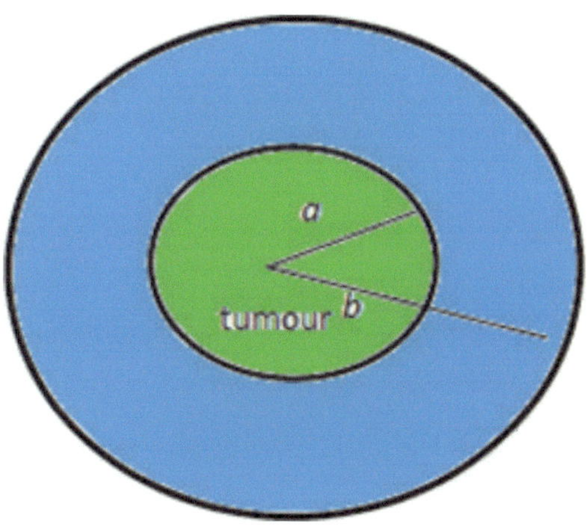

Figure 1.
Tumor tissue, spherical in shape, surrounded by normal tissue concentric sphere [20].

$$T_1(0,t)\ \text{isfinite} \tag{5}$$

$$T_1(r,0) = T_0 \tag{6}$$

$$T_2(r,0) = T_0 \tag{7}$$

To solve the above system of differential equations, we discretize it using the Euler method, which is reproduced here [20].

Let h > 0 and k > 0 be the step lengths in the space and time directions, respectively. Also let N1 and N2 be integers such that

$$\text{hN1} = \text{a and hN 2} = \text{b}. \tag{7a}$$

We replace the region $\Omega = \{(r,t)\ 1\ 0 \le r \le b, t \ge 0\}$ by a set of grid points (r_1, t_j), denoted by (l, j),

where

$$r_1 = lh, t_j = jk, j = 0, 1, 2, \ldots, J \text{ and } l = 0, 1, \ldots N_2. \tag{7b}$$

where J is a positive integer.

Let

$$(T_1)_l^j = T_1(r_l, t_j) \text{ and } (T_2)_l^j = T_2(r_l, t_j) \tag{7c}$$

denote the solution of (1) and (2), respectively, at the grid point (l, j).

We approximate the solution of (1), (2) at the grid point (l, j) by the scheme

$$(T_1)_l^{j+1} - (T_1)_l^j = \frac{k_1 k}{c_1 \rho_1 h^2}\left[\left((T_1)_{l+1}^{j+1} - (T_1)_l^{j+1}\right) - \left(1 - \frac{2k}{r}\right)\left((T_1)_l^{j+1} - (T_1)_{l-1}^{j+1}\right)\right] + \frac{Pk}{c_1 \rho_1}, \quad 0 < l < N_1 \tag{8}$$

$$(T_2)_l^{j+1} - (T_2)_l^j = \frac{k_2 k}{c_2 \rho_2 h^2}\left[\left((T_2)_{l+1}^{j+1} - (T_2)_l^{j+1}\right) - \left(1 - \frac{2k}{r}\right)\left((T_2)_l^{j+1} - (T_2)_{l-1}^{j+1}\right)\right] + \frac{Pk}{c_2 \rho_2}, \quad N_1 < l < N_2 \tag{9}$$

$$(T_1)_0^{j+1} - (T_1)_0^j = \frac{k_1 k}{c_1 \rho_1 h^2}\left((T_1)_1^j - (T_1)_0^j\right) + \frac{Pk}{c_1 \rho_1}, \tag{10}$$

From the boundary condition at the tumor-healthy tissue edge, we can write the approximation at the grid point (N_1, j) as

Parameter	Constant	Value
Radius of liver tumor	A	2.50 cm
Tumorous liver tissue specific heat	c_1	3.758 kJ/kg K
Healthy liver tissue specific heat	c_2	3.617 kJ/kg K
Liver tissue heat conductance	k_1=k_2	0.5122 W/m K
Liver density	$\rho_1 = \rho_2$	1.0492 g/mL

Table 1.
Liver tissue and nanoparticle parameters [11, 24–26].

$$(T_1)_{N_1}^{j} = (T_2)_{N_1}^{j} = \frac{k_1 (T_1)_{N_1-1}^{j} + k_2 (T_2)_{N_1}^{j}}{k_1 + k_2} \tag{11}$$

Solving Eqs. (8)–(11), using values of the different constants from **Table 1** [11, 24–26], the graphs can be plotted for studying dependency of temperature in and around tumor on radius (of tumor), time (of heat exposure), and power applied (on magnetic nanoparticles).

4. Results and discussion

From **Table 1**, it can be observed that varying r from center (0 cm) to the boundary of the affected area (=a), time of exposure up to 10 s with power ranging

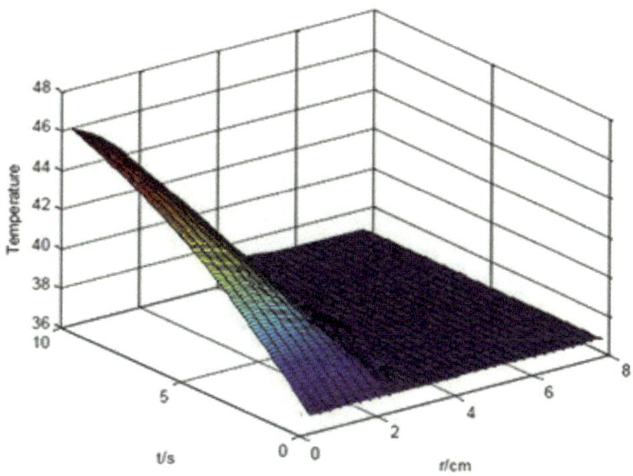

Figure 2.
For a constant magnetic nanoparticle power of 5 W, temperature in the tumorous cell/tissue plotted, as a surface plot and as a function of hyperthermia time (t, in s) and distance from the center of the tumor (r, in cm) [20].

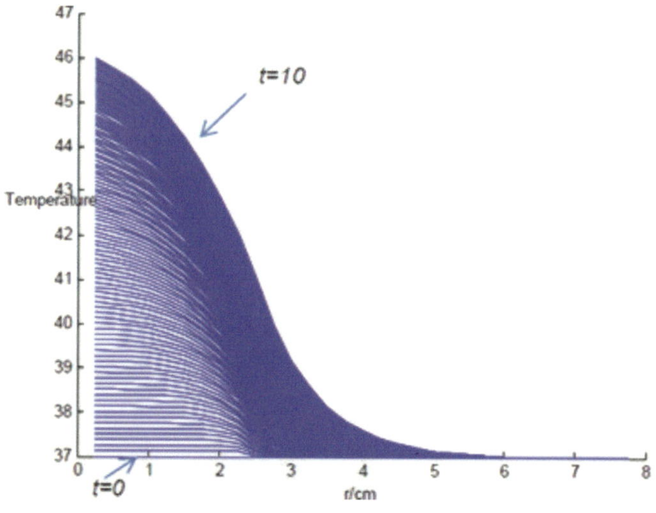

Figure 3.
For a constant magnetic nanoparticle power of 5 W, plot of temperature inside the tumorous cell/tissue as a function of distance from the center of the tumor (r in cm) and hyperthermia time (t, in s) [20].

from 2.75 to 6.5 W, MATLAB software is used to obtain **Figures 2–6**. In the given model, it was assumed that magnetic nanoparticles used were of the size up to 10 nm in radius. The surface plot (**Figure 2**) shows dependency of temperature in the affected area on hyperthermia time (t in s) and radius of the tumor (r in cm). It can be concluded from it that on application of 5 W power, temperature in tumor increases to 46°C at the middle region of the tumor, gradually reducing to body temperature at the interface (of affected and healthy area), thus causing least effect to the unaffected area. Similar conclusion can be drawn from **Figure 3**. The temperature variation at the center of the tumor on varying power of magnetic nanoparticles and time of heat exposure can be studied from **Figures 4** and **5**. It is observed that the temperature at the middle of the tumor cell gradually increases from body temperature at the interface (r=a) to 48°C for an applied power range of 2.75–6.5 W. It can be further observed that for a standard time of exposure (t = 7 s), if the power is varied from 2.75 to 6.5 W over a radius of 2.5 cm, it leads to annihilation of the tumor, from **Figure 6**.

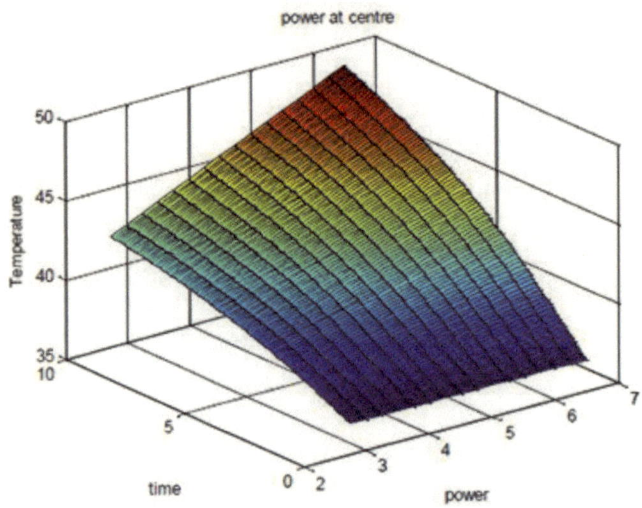

Figure 4.
Temperature at the center of the tumorous cell/tissue as a function of hyperthermia time (t in s) and a magnetic nanoparticle power (p, in Wts) [20].

Figure 5.
Temperature at the center of the tumorous cell/tissue plotted as a function of magnetic nanoparticle power (p, in W) and hyperthermia time (t, in s) [20].

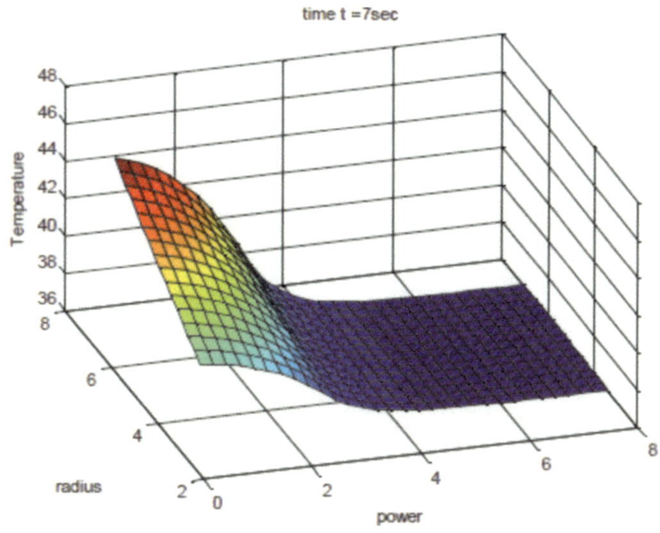

Figure 6.
Temperature in the tumorous cell/tissue plotted as a function of distance from the center of the tumor (r, in cm) and magnetic nanoparticle power (p, in watts) for a constant exposure time (t = 7 s) [20].

5. Conclusion

From the discussion, it can be concluded that hyperthermia treatment involving magnetic nanoparticles can be efficiently and effectively used for the removal of tumorous cell/tissue with not much collateral damage. Reduced damage to the neighboring healthy cells makes the technique more successful, clinical results are also in tandem with results if we use nanoparticles with power in the range of 2.75–6.5 W with a heat exposure time up to 10 s, this futuristic approach will make treatment more effective with fewer side effects and less cost, leading to widespread use and finally conquering the disease, which even in this robotic age is considered a taboo.

As recently engineers from MIT are working on designing of tiny robots, in nano range, which can assist in drug delivery, the engineered robots called the "microbots" are based on bacterial propulsion. The scientists have proposed that the procedure can help in overcoming the hindrances to drug delivery loaded with nanoparticles, enabling them to exit blood vessels and hit the right place [27].

As recently as March 2019, Angl Apostolov et al. reported the study based on similar concept of destroying hyperthermia. The group suggests, theoretically, the use of mixed ferrite nanoparticles with structure formula $Me_{1-x}Zn_xFe_2O_4$ (Me = Co, Ni, Cu, Mn) appropriated for self-controlled magnetic hyperthermia (SMHT) for both in vivo and in vitro applications.

Thus, there is a lot of scope in the area which can be reinvented and researched to make the world if not free at least less cancer deaths or sufferings.

Author details

Neha Sharma
University of Delhi, Delhi, India

*Address all correspondence to: nehasharma2710@hotmail.com

References

[1] Jemal A et. al. Cancer ststistics. Cancer Journal of Clinicians. 2007;**57**(43)

[2] Cancer Reference Information. American Cancer Society, Inc.; 2006

[3] Arruebo M, Pacheco RF, Ibarra MR, Santamaria J. Magnetic nanoparticles for drug delivery. Nanotoday. **2007**;**2**(3): 22-32

[4] Brigger I. Advanced Drug Delivery Reviews. 1999;**37**(121)

[5] Jurgons R et al. Drug loaded magnetic nanoparticles for cancer therapy. Journal of Physics: Condensed Matter. 2006;**18**(S2893)

[6] Bogdanov A et al. A long-circulating co-polymer in 'passive targeting' to solid tumours. Journal of Drug Targeting. 1997;**4**(321)

[7] Ritter JA. Journal of Magnetism and Magnetic Materials. 2004;**280**(184)

[8] Strebhardt K, Ullrich A. Paul Ehrlich's magic bullet concept: 100 years of progress. Nature Reviews. Cancer. 2008. Advanced Online Publications

[9] Britt BA, Kalow W. Malignant hyperthermia: A statistical review. Canadian Anaesthetists Society Journal. 1970;**17**(4):293-315

[10] Dewey WC, Hopwood LE, Sapareto SA, Gerweck LE. Cellular responses to combinations of hyperthermia and radiation. Radiology. 1977;**123**(2):463-474

[11] Andrä W, d'Ambly CG, Hergt R, Hilge I, Kaiser WA. Temperature distribution as function of time around a small spherical heat source of local magnetic hyperthermia. Journal of Magnetism and Magnetic Materials. 1999;**19**:197-203

[12] Ivkov R, DeNardo S. Application of high amplitude alternating magnetic fields for heat induction of nanoparticles localized in cancer. Clinical Cancer Research. 2005;**11**(19 Suppl):7093s

[13] Goya GF, Grazu V, Ibara MR. Magnetic nanoparticles for cancer therapy. Current Nanoscience. 2008;**4**(1)

[14] Lin C, Liu K. Estimation for the heating effect of magnetic nanoparticles in perfused tissues. International Communications in Heat and Mass Transfer. 2009;**36**:241-244

[15] Wappler F. Malignant hyperthermia: Current strategies for effective diagnosis and management

[16] Expert Opinion on Orphan Drugs. 2014;**2**(3):259-269

[17] Freeman MW. Magnetism in Medicine. Journal of Applied Physics. 1960;**31**(S404)

[18] Polyak B, Friedman G. Magnetic targeting for site-specific drug delivery: Applications and clinical potential. Expert Opinion on Drug Delivery. 2009; **6**(1):53-70

[19] Tsafnat N. Modelling heating of liver tumors with heterogeneous magnetic microspheres disposition. Physics in Medicine and Biology. 2005; **50**:2937-2953

[20] Schuerle S et al. Robotically controlled microprey to resolve initial attack modes preceding phagocytosis. Science Robotics. 2017;**2**(2)

[21] Sharma N, Singh S, Singh S. Optimizing the power required in hyperthermia treatment using magnetic nanoparticles. International Journal of Control and Automation. 2016;**9**(9): 181-188

[22] Wu J. Hyperthermia cancer therapy by magnetic nanoparticles. Available from: www.isn.ucsd.edu/classes/beng 221/problems/2013/project-9

[23] Carslaw HS, Jaeger JC. Conduction of Heat in Solids. Oxford: Clarendon Press; 1967

[24] Valvano JW, Cochran JR, Diller KR. Thermal conductivity and diffusivity of biomaterials measured with self-heated thermistors. International Journal of Thermophysics. 1985;**6**(3):301-311

[25] Howells CC. Normal liver tissue density dose response in patients treated with stereotactic body radiation therapy for liver metastases. International Journal of Radiation Oncology Biology Physics. 2012;**84**(3):e441-e446

[26] Giering K, Lamprecht I, Minet O. Determination of the specific heat capacity of healthy and tumorous human tissue. Thermochimica Acta. 1995;**251**(1):199-205

[27] Apostolov A, Apostolova I, Wesselinowa J. Specific absorption rate in Zn-doted ferrites for self-controlled magnetic hyperthermia. The European Physical Journal B. 2019;**92**:58. DOI: 10.1140/epjb/e2019-90567-2

Chapter 10

Chemical Synthesis and Characterization of Luminescent Iron Oxide Nanoparticles and Their Biomedical Applications

Martin Onani, Leandre Brandt and Zuraan Paulsen

Abstract

The syntheses and characterizations of biocompatible luminescent magnetic iron oxide nanoparticles has drawn particular attention as diagnostic and drug delivery tools for treatment of cancer and many other diseases. This chapter focuses on the chemical synthetic methods, magnetic and luminescent properties, including the biomedical applications of iron oxide nanomaterials and luminescent magnetic iron oxide-based nanocomposite materials. The influences of functionalizing with short ligands such as dopamine and L-cysteine on the magnetic properties of synthesized nanoparticles are described. The chapter contains some data on necessary reagents and protocols for bioconjugation aimed at cell culture and step by step the MTT assays used to evaluate cytotoxicity are also presented. In the final section of the chapter, we focus on the biomedical applications specifically for diagnosis and treatment of breast cancer treatment. This chapter also investigates the application of various characterization techniques for analysis of the structural, optical and magnetic properties of the iron oxide nanoparticles and as their nanocomposites.

Keywords: nanoparticles, quantum dots, nanocomposite, magnetic, luminescence, biomedical, imaging, cytotoxicity, diagnostics

1. Introduction

Over the past few decades, magnetic nanoparticles (MNPs) have attracted tremendous attention due to their unique and tunable chemical and physical properties. Magnetic nanoparticles can selectively target specific biological surfaces of interest owing to the arrangement in dipoles in the absence and presence of an external magnetic field. Iron oxide nanoparticles (IONPs) are one of the mostly used MNPs since they are nontoxic and biodegradable, being promising candidates for use in biology and medicine for example imaging [1, 2], siRNA and drug delivery [3, 4], cell tracking [5], magnetic separation [6, 7] hyperthermia [8, 9], and bio- and chemo-sensoring and [10] biomedical applications. Moreover, IONPs are mostly used as magnetic resonance imaging (MRI) probes to differentiate between normal and cancerous cells for diagnosis [11, 12]. Moreover, magnetic imaging has no practical depth limitation for imaging, however spatial resolution is poor and multiple imaging is problematic. To improve the versatility and efficiency in numerous technologies, the development of hybrid magnetic nanoparticles combining both fluorescent and magnetic properties magnetic are being developed [13–23]. The combination of MRI and fluorescent spectroscopy in one nanocomposite opens up unique multimodal properties to monitor complementary information in biological applications such as in multimodal biological imaging, drug delivery systems and medical diagnostics. Despite many problems related to the synthesis of hybrid magnetic-fluorescent nanoparticles, major advances in recent years have been made in this field. For the synthesis, both physical and chemical techniques have been used for the synthesis of IONPs; still, the chemical approach are easier to control the NPs, such as the co-precipitation, thermal decomposition, hydrothermal synthesis, microemulsion, and sol-gel and polyol methods. Of all these approaches, the chemical approach, particularly co-precipitation method is discussed in Section 2. As Fe_3O_4 NPs are the mostly used IONPs, in this section we focus on the chemical synthesis of Fe_3O_4 NPs. Also, covered in this section is the synthesis of fluorescent-magnetic nanocomposite material, using InP/ZnSe NPs as fluorophore. The syntheses of fluorescent-magnetic nanoparticles are challenging due to chemical stability and the aggregation of the nanoparticles in solution caused by electron transfer interactions between the particles. The main challenge associated is to overcome the quenching of the luminescence of the fluorophore when it is on the particle surface of the magnetic core. This can be due to the electron and energy transfer between the fluorophore and the magnetic nanoparticles [24–26]. The easiest and most commonly used method to overcome this hurdle is to isolate the magnetic core from the fluorescent molecule. This can be achieved by coating the magnetic nanoparticle with a shell before it is attached to the fluorescent structure or by placing a spacer between the two molecules. These solutions lead to most luminescent magnetic nanoparticles to have a core-shell structure [15]. The shell needs to have specific properties namely: non-toxic or harmful to human tissue, should not cause the body to emit an immune response, to avert or reduce agglomeration and reduce non-specific interactions with proteins, cells and other components of biological media. Hence, Section 3 covers several procedures for the functionalization and formation of the fluorescent-magnetic nanocomposite material to overcome these challenges. In Section 4, the biomedical applications of IONPs including MRI, magnetic hyperthermia, magnetic targeting, and cell tracking, with focus on diagnosis for breast cancer treatment are reviewed.

1.1 Purpose of the study

Nanocomposite material with dual or multiple properties have shown extensive potential to improve the performance of current cancer diagnostic tools and/or therapy, for biosensor applications, *in vivo* optical imaging or drug delivery. The aim of this project is to synthesize a nanohybrid material with luminescent and magnetic properties and having low or no toxicity, to be used for biological studies.

In this experiment the synthesis of the multifunctional material will be synthesized via the process seen in **Figure 1**. From the diagram the end product, the nanocomposite material, the QDs are expected to cluster around the MNPs.

In order to synthesize the Fe_3O_4-InP/ZnSe bifunctional nanocomposite material, the luminescent InP/ZnSe nanocrystals were prepared separately from the Fe_3O_4 magnetic nanoparticles. Once both the MNPs and QDs nanomaterials are synthesized they are both will be functionalized with a compound containing a thiol group. The MNPs and QDs were functionalized with dimercaptosuccinic acid (DMSA) and mercaptopropionic acid (MPA), respectively. Using thiol chemistry, the QDs will directly combine to the surface of the MNPs (as seen in **Figure 1**).

DOI: http://dx.doi.org/10.5772/intechopen.88165

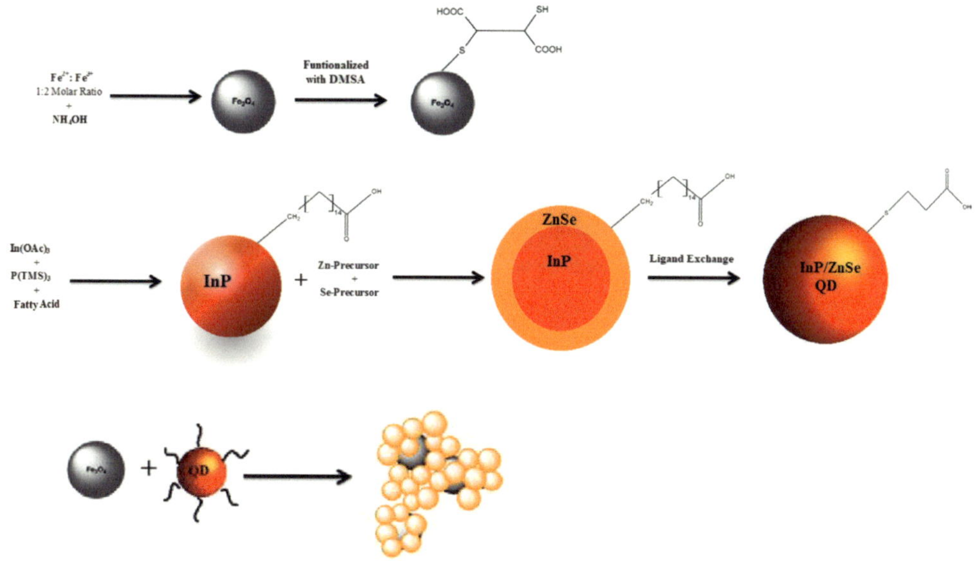

Figure 1.
Experimental schematic of the synthesis of the magnetic-luminescent nanomaterial.

2. Chemical synthesis of magnetic nanoparticles and magnetic-fluorescent nanoparticles

2.1 Magnetic nanoparticles: synthesis

In this chapter, we discuss the general and recent progress of different chemical synthetic pathways for IONPs (Fe_3O_4). Their small and controllable sizes, easily functionalized, as well as the ability to be manipulated by external magnetic forces [15], are all attractive properties for various applications including biomedical pursuits. The properties of MNPs strongly depended on the synthesis route. Consequently, the controllable synthesis of monodispersed IONPs is critical for controlling their size distribution, structural defects, surface chemistry, and magnetic behavior for application in specific biomedical field. The synthesis of shape-controlled, stable, biocompatible, and monodispersed IONPs have drawn much effort over recent years. IONPs have been produced by various chemical, physical and biological methods which have both advantages and disadvantages (**Table 1**). Chemical synthesis offers significant advantages over other methods, as it is a facile, cost-effective method with ease of control over the NPs characteristics. These methods include thermal decomposition, co-precipitation, microemulsion, hydrothermal synthesis, and sol-gel and polyol methods, also shown in **Table 1** [32]. Of these methods, co-precipitation is the mostly used as it tends to be green, simple and effective with low production cost, high reproducibility and high yields in one synthesis [27]. Hence, it is of interest, and discussed in detail in section below.

2.1.1 Co-precipitation method

Co-precipitation method is the preferred choice among studied synthetic methods for the preparation of Fe_3O_4 NPs. It is a simple and classical approach to follow as it is simple, convenient, cheap with high reproducibility, solubility and scalability for large scale production. However, due to the high influence of kinetic factors on the growth of Fe_3O_4 NPs, such as low reaction temperatures, this resulted

Techniques	Advantages	Drawbacks
Chemical		
Microemulsion	Precise control over shape and size and high surface area-to-volume ratio	Complex, low yields
Co-precipitation	Simple, convenient, very effective, cheap, standard ambient conditions	Broad size distribution and poor shape control
Sonochemical	Easy with narrow size distribution	No control over shape and medium yield
Hydrothermal	Ease of control over size and shape. High efficiency	High pressure and high reaction temperature
Polyol and sol-gel	Facile with precise control over size and their internal structure	Complicated and high pressures
Electrochemical decomposition	Good control over particle size	Very low reproducibility, rough and amorphous impurities final product
Thermal decomposition	Monodispersed NPs, excellent shape and size control	Only dissolves in non-polar solvents
Physical		
Electron beam lithography	Good control over inter-particle spacing	Costly and use extremely complex apparatus
Gas phase deposition	Easy	Poor control over size
Aerosol	Relatively narrow size range	Complex
Biological		
Bacteria-mediated	Cheap, good reproducibility and scalability, and high yields	Tedious, laborious

Data edited from [27–31].

Table 1.
A comparison of the several synthetic methods of IONPs with advantages and disadvantages.

in formation of irregularly shaped NPs with broad size distribution. This is the best method for the synthesis of water soluble magnetic nanoparticles. However, it has major drawbacks of broad particle size distribution [27, 33]. In 1982, Rene Massart prepared the first superparamagnetic iron oxide NPs, magnetite (Fe_3O_4), [17] via an alkaline precipitation of $FeCl_3$ and $FeCl_2$ mixture in a molar ratio of 1:2 [17]. The NPs were some-like spherical shaped, with diameter in broad size range around 8 nm. Hence, a size selection process using NaCl as an extra electrolyte was used to selectively decrease the electrostatic repulsions between NPs. This caused aggregation and formation of larger colloidal particles in the supernatant with a diameter of about 7 nm [18]. Hence, O_2-bearing atmospheres is required for subsequent reactions to form maghemite (Fe_2O_3) or ferric hydroxide ($Fe(OH)_3$) [27, 30, 31], due to the sensitivity and instability of magnetite as it is prone to oxidation [34]. In most cases, the co-precipitation method involves some form of mixing Fe^{2+}/Fe^{3+} salt solutions in an alkaline medium at standard or elevated temperatures under inert (N_2 or Ar) atmospheres to avoid the possible oxidation of Fe^{2+} into Fe^{3+} [35]. Most papers apply temperatures between 60 and 80°C, some at even higher temperatures [36]. The alkaline solutions commonly used are sodium hydroxide, potassium hydroxide and ammonium hydroxide. The co-precipitation method consists of two major steps—the first is the occurrence of a short nucleation burst at critical supersaturation is reached, and the second involves the slow nucleation growth via diffusion of the solute to the nanocrystal [15]. To obtain monodispersed Fe_3O_4 NPs, these two stages must be kept separate and the Fe^{2+}/Fe^{3+} must be fixed at 1:2 molar ratio. Large amounts of monodispersed IONPs can be easily synthesized by changing certain reaction parameters for example the pH, temperature, ionic strength, composition of iron salts, ratio of ferrous to ferric iron and the type of the base and salt precursors [27]. Depending on the parameters, the particle size can be tuned in size range of 2–15 nm [27] with superparamagnetic properties. In most cases, the particle size increase as reaction time and temperate increase, the faster reaction rate results in formation of monomeric generation of NPs. Moreover, the pH value has shown an important role in controlling the size and stability [16]. Studies have shown that the pH must be kept in the range of 8–14 for monodispersed IONPs. A decrease in the pH value results in the decrease of the diameter or dissolution of the NPs, while increase in pH value show NP tendency to oxidate.

This can be prevented by using a surfactant on the surface of Fe_3O_4 NPs which cause repulsive force between radical ions. In addition, the surfactant not only protects the surface of Fe_3O_4 NPs, but can also control the size of NPs. In a paper Gao et al. synthesized Fe_3O_4 NPs using an aqueous solution of $FeSO_4.4H_2O$, $NaNO_3$, NaOH, and citrate as the surfactant [19]. The diameter range was tuned from 20 to 40 nm by changing Fe^{2+} concentration. In another paper, Kumar et al. report on an environmentally benign, non-toxic and cost-effective method for the success synthesis of spherical shaped Fe_3O_4 NPs.

Blackberry leaf (ABL) extract is used as capping agent, and added to a solution of $FeSO_4.7H_2O$, using NaOH to adjust the pH to 10–11, the solution was gradually heated between 75 and 80°C. The obtained Fe_3O_4 NPs had a size range of 54.5 ± 24.6 nm diameter [40]. In our group, Kiplagat et al. synthesized bare and meso-2,3-dimercaptosuccinic acid (DMSA) capped Fe_3O_4 (as shown in **Figure 2**). We prepared bare iron oxide nanoparticles following a simple co-precipitation method by dissolving salts of Fe^{3+} and Fe^{2+} with a molar ratio of 1:2 at the pH of

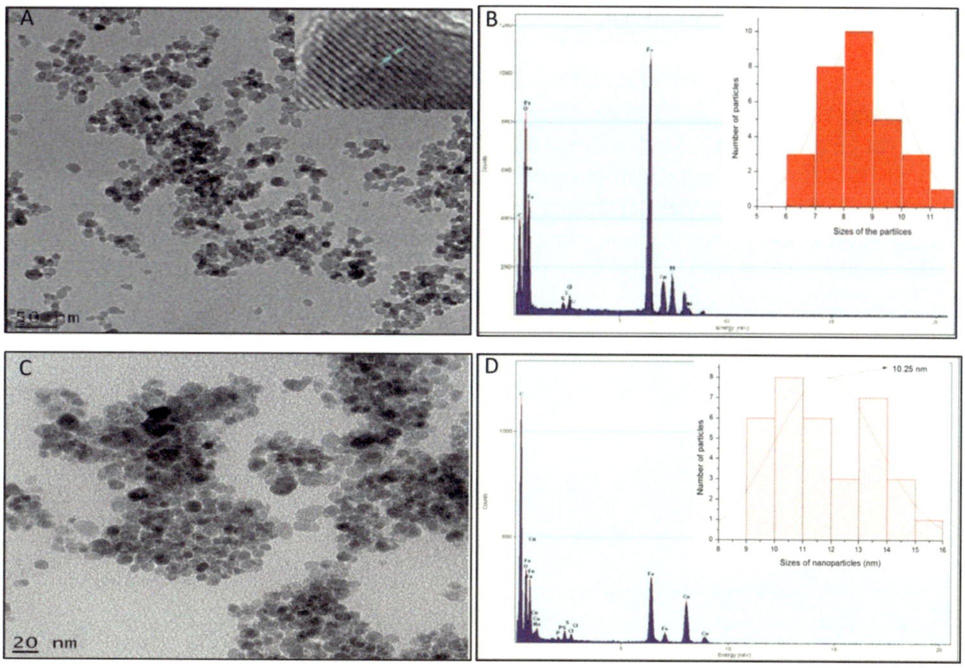

Figure 2.
TEM and EDX of bare and DMSA capped irregularly spherical shaped Fe_3O_4 nanoparticles synthesized by co-precipitation method (insets: histograms showing size distribution).

12 in ultra-pure water at 50°C [41]. The DMSA capped iron oxide nanoparticles were prepared by dispersing IONPs in toluene and dimethylsulfoxide solution. The diameter of the bare IONPs range from 6 to 13 nm and the average size of 8.5 nm, whereas the DMSA capped IONPs size distribution range from 9 to 16 nm with an average diameter of 10.25 nm as shown histograms in **Figure 2**. The average size of DMSA capped iron oxide nanoparticles increased slightly compared to the bare iron oxide nanoparticles. Consequently, in our study we found that the synthesis of highly crystalline and monodispersed Fe_3O_4 NPs was not as easy to achieve. The capping of the iron oxide nanoparticles with the DMSA resulted to partial agglomeration as seen **Figure 2(a)** and **(c)**. We speculate that the presence of two thiol functional groups in DMSA lead to coupling the nanoparticles, thus the tendency to agglomerate. Similar observations were made by Kumar et al. [40] where they noted that functionalization with ABL led to partial aggregation and broader particle size distribution of IONPs. The use of suitable capping ligands is a widely used to improve their biocompatibility and stability [27, 44], however various approaches are being employed to avoid aggregation of these magnetic nanoparticles that restricts their applications.

In **Figure 3**, we show the XRD patterns of bare and DMSA capped iron oxide nanoparticles. The XRD pattern was matched with JCPDS no. 00-039-1346 and 00-019-0629 for maghemite (α-Fe_2O_3) and magnetite (Fe_3O_4), respectively. Our iron oxide nanoparticles were found to be maghemite. This might be due to the IONPS tendency to undergo oxidation as mentioned earlier [27, 30, 31, 35, 36]. A comparison chart of different coated IONPs and nanocomposites, including their characteristics such as the capping agent, physicochemical and magnetometric properties are shown in **Table 2**. The main of objective was to provide a facile method of preparing Fe_3O_4 NPs and its corresponding nanocomposites, to overcome the drawbacks of the prior research, particularly a process with less chemical reagents, carried out at standard reaction conditions.

2.2 Fluorescent QDs: synthesis

On the other hand, another class of nanomaterials is quantum dots. Quantum dots (QDs) are inorganic fluorescent semiconductor nanoparticles with dimensions in the range of 1–10 nm. The QDs are usually composed of atoms from groups II–VI, III–V, or IV–VI [45]. The nanometer dimensions of the QDs causes a confinement of electron and hole carriers at dimensions smaller than the bulk Bohr excitation radius; this causes phenomenon, called quantum confinement, to occur in these

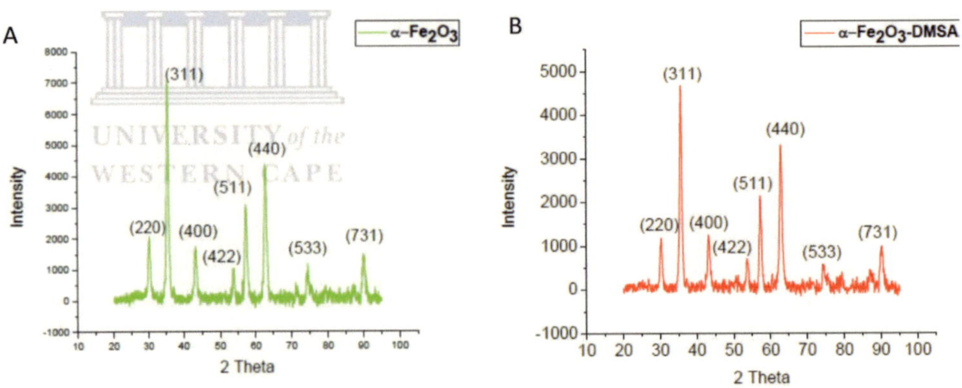

Figure 3.
XRD pattern for bare and DMSA capped iron oxide nanoparticles.

Type of IONPs	Synthetic method	Coating	Physicochemical properties	Magnetometric properties (SQUID)	Application	Reference
$MnFe_2O_4$	Thermal decomposition	1,2-Hexadecanediol, dodecanoic acid, dodecylamine	Spherical shape; 12 nm	M_s = 298 K, T_b = 43.2 emu/g (SQUID)	Multimode imaging probes	Lim et al. [26]
Fe_3O_4	Thermal decomposition of iron oleate in NaCl	Oleic acid	Octapodes; 20–30 nm	M_s = 51–71 emu/g, T_b = 240–290 K (SQUID)	Magnetic resonance imaging	Zhao et al. [37]
Fe_3O_4	Co-precipitation	DMSA	Nearly spherical, 1.4–6.5 nm	M_s = 51.7 emu/g, T_b = 298 K (VSM)	Lymphoma treatment	Song et al. [38, 39]
Fe_3O_4	Green method	Andean blackberry leaf (ABL) extract	Spherical shape, 54.5 ± 24.6 nm	N/A	Degradation of organic dyes, antioxidant	Kumar et al. [40]
Fe_3O_4	Co-precipitation	DMSA	Spherical shape, 9–16 nm	M_s = 43.2 emu/g (SQUID)	Breast cancer treatment	Kiplagat et al. [41]
InP-Fe_3O_4	Co-precipitation	DMSA	Agglomerated, unable to obtain particles	M_s = 6.03 emu/g (SQUID)	Breast cancer treatment	Kiplagat et al. [41]
Core-shell Fe_3O_4@SiO_2@Au	Thermal decomposition, sol-gel and oil-in-water technique	Cetyltrimethylammonium bromide (CTAB)	Spherical shape, 100–110 nm	M_s = 50–46 emu/g, T_b = 298 K (SQUID)	Photoacoustic and magnetic resonance imaging detection	Monaco et al. [42]
Fe_mO_n-SiO_2	Co-precipitation	Silica	Irregular nanoflakes; 98–101 nm	M_s = 11 emu g^{-1} (VSM)	Cytotoxicity	Toropova et al. [43]
SiO_2-FemOn	Co-precipitation	Silica	Core-shell structure; 98–101 nm	M_s = 37.2 emu g^{-1} (VSM)	Cytotoxicity	Toropova et al. [43]

M_s = saturation magnetization; T_b = blocking temperature; D_c = critical size; and H_c = coercivity field.

Table 2.
A comparison of several methods of the organic and inorganic coated IONPs, and their corresponding characteristics.

nanoparticles. The QDs have tunable energy, optical and electronic properties which are done by either managing the QDs size or composition. QDs may be produced via various methods. These methods include but not limited to colloidal synthesis, plasma synthesis, self-assembly, and electrochemical assembly. However, to be able to tune the QDs to a have desired properties and to produce high quality QDs colloidal synthetic methods are the easiest and the most explored. Since there are several different types of QDs in exists, for simplicities sake this study will focus on InP.

A colloidal synthetic approach to manufacture QDs can either be achieved by a heating-up technique or rapid hot injection method. The heating up method is a batch process and is achieved by adding all the desired chemicals to a reaction vessel at relatively low temperatures or at room temperature, followed by rapidly heating the entire reaction up to a desired temperature that allows for crystal growth to occur. Khanna et al. [46] directly synthesized indium phosphide (InP) nanoparticles by heating a solution of indium powder in n-trioctylphosphine (TOP). The reaction was carried out under an argon atmosphere. The raw materials, the reaction time and temperature were varied to determine which reaction conditions would create the finest results. In addition, their research demonstrated that with high temperatures in conclusion with, short reaction times and a low amount of TOP leads to InP nanoparticles with small particle sizes and less impurity. The formation of the InP is caused by the catalytic activity of indium nanoparticles attempting to reduce C-P bonds found in TOP. The synthesis method is considered to be simple, low cost and avoids the use of hazardous and expensive raw materials.

2.2.1 Hot-injection method

To obtain QDs via the rapid hot injection technique, a main reaction is heated to a desired temperature and room temperature precursors are added to the reaction by rapidly being injected into the reaction. The quick addition of the precursors causes the reaction to supersaturate thus allowing for nucleation to occur. The reaction temperature when the cooler precursor is added, the addition of the precursor also causes the reaction to become diluted. The lowered temperature and the lowered concentration of the reaction materials prevent further nucleation, but nanocrystal growth still occurs. The work of Lui and co-workers used a reduction colloidal approach Lui and co-workers [47] were able to synthesize high quality InP NCs. The synthesis required the use of octadecene as a solvent and stearic acid as a capping ligand. These were heated in the presence of indium acetate, under an inert atmosphere for 30 min. After being heated a PCl_3-precursor was added to the solution at 40°C. The temperature was then elevated to allow the growth of the InP core. By varying the reaction time and the temperature the study showed how the NC growth could be tuned to a desired size and size distribution. Upon further investigation it was demonstrated that, using a HF post-production treatment, the photoluminescence could be vastly enhanced. The research conducted confirms that InP NCs can be synthesized without the use of the hazardous and expensive material $P(TMS)_3$.

2.3 Synthesis of IONPs/QD hybrid nanocomposite

Creating a complex system like this presents a few complications. The fabrication of such a system would require several synthesis and purifications steps, which is time consuming and expensive. Also having a magnetic element in the presence of a fluorescent compound reduces the photoluminescence and could quench the fluorophore [47–50]. There is currently very little information for possible alternatives to the first problem; the second is usually solved by encasing the MNP's in silica or polymer [24]. In the research conducted by Hong et al. [50], the layer-by-layer (LbL) approach was used to synthesize the magnetic-luminescent nanocomposite. LbL approach is based on the electrostatic attraction of oppositely charged species. In this synthesis, MNPs were used as a template for the multiple deposition of CdTe QDs. The MNPs were synthesized using the co-precipitation of ferric chloride and ferrous chloride. While TGA-capped CdTe QDs were prepared by the addition of Cd^{2+} into a solution of NaHTe in the presence of TGA. Using LbL they were able to fabricate $Fe_3O_4/PE_n/CdTe$ and $Fe_3O_4(PE_3/CdTe)n$ nanocomposite material by varying the number of deposition cycles of polyelectrolyte layers. The polyelectrolyte phases allowed for increased PL intensity while maintaining strong magnetic properties.

The synthesis described by Gua et al. [51], demonstrated the synthesis of a multifunctional system by integrated materials, with luminescent and magnetic properties, into microspheres of quantum dots (QD) with a cross-linked polymer shell. They basically synthesized iron oxide magnetic nanoparticles (MNP) via a co-precipitation method and thiol-capped cadmium telluride (CdTe) by hydrothermal route. The MNP were incorporated into a silica sphere via the Stöber method and the QDs added. These conjugated moieties were capped using a template polymerization. Their technique provides many advantages including the formation of a robust luminescent shell with multicolor bar codes which is generated by the aggregation of the thiol-capped CdTe on the silica particles. The outer shell not only protects the CdTe shells from damage, but also facilitates the covalent bonding of the ligands to the nanoparticles.

The synthesis described by Liu et al. [48], demonstrates the synthesis of a magnetic-luminescent MNP-QD nanocomposite via electrostatic interactions. The two major problems that occur is (i) the close interactions of the QDs and MNPs, when they are embedded in a matrix, material causes photobleaching; (ii) while the layer-by-layer process takes an extremely long time and a lot of effort. In this experiment first CdSe QDs are synthesized via the hydrothermal route. While the MNPs are separately prepared using the co-precipitation of Fe^{2+} and Fe^{3+} salts, followed by a silica coating by means of the Stöber method, and finally functionalizing the silica coated MNPs with 3-aminopropyltrimethoxysilane (APTS). The final MNPs were added to a solution of CdSe QDs. The new solution was sonicated and stirred for 6 hours at room temperature. The nanocomposite material was collected and separated by magnetic decantation. The luminescent-magnetic nanomaterial was spherical and had a diameter between 95 and 105 nm. The MNP-QD interactions caused a decrease in the PL intensity. The M_s of the silica-MNPs was 5.4 emu g^{-1} and the 3.8 emu g^{-1} for the nanocomposite.

The increase attention of multifunctional nanomaterial has led Nai-Qiang et al. [52] to develop a method to synthesize a nanocomposite composed of MNPs and QD material. The synthesis started the QDs and MNPs were prepared separately. The Mn-doped ZnS QDs synthesis began with a solution of $Zn(NO_3)_2$, manganese acetate and 3-mercaptopropionic acid (MPA) being mixed together. After the mixture undergone a dilution, an adjustment of the pH, and purging the air with N_2, NaS_2 was injected into the solution. The MNPs were synthesized via the co-precipitation of $FeCl_3$ and $FeSO_4$. The MNPs were then coated with SiO_2 using the Stöber method, finally the coated MNPs were modified with APTS. Using electrostatic interactions, the MNP-QD linkage was able to occur after 6 hours of rapid stirring. The nanocomposite material XRD pattern was a combination of the SiO_2-MNP and Mn-doped QD patterns. The TEM results measured the nanocomposite material to be 100–130 nm in size and spherical. The SQUID analysis showed a decrease of from 54 to 7 emu g^{-1} once the MNPs were coated with SiO_2 and the nanocomposite was also roughly 7 emu g^{-1}.

Due to the great potential surrounding multifunctional nanomaterial, there is a desire to create a fast, simple and large-scale synthesis of the nanocomposite material. Microwave irradiation (MWI) been successful in synthesizing various nanostructures that Zedan et al. [53] attempted to use the design to develop a novel synthesis of the magnetic-luminescent nanocomposite material. Using microwave synthesis, the Fe_3O_4 and CdSe NCs were prepared separately and to create the nanocomposite material a seed-mediated approach was used. The Fe_3O_4 NCs were used as seeds and CdSe semiconductor material was allowed to grow around the nanoparticle under MWI. The TEM images of the nanocomposite material confirmed that they maintained the core-shell morphology, were spherical and 10–15 nm in size. The XRD pattern of the nanocomposite material showed the material having good crystallinity. The nanocomposite material maintained the same emission and adsorption peaks as the CdSe QDs. Also changing the irradiation time provides the nanocomposite material with tunable optical properties and the ability to control the luminescent shell's thickness.

As mentioned previously the synthesis of such a material is very complex, Cho et al. [54] tried to optimize the synthesis of MNP-QD hybrid system by using a direct nucleation route. The multifunctional nanomaterial was prepared by first synthesizing the iron oxide NCs via the thermal decomposition of FeO(OH) with oleic acid as a surfactant in octadecene. The MNPs which formed were then purified and dispersed in hexane. The synthesis of the complex was created during the synthesis of cadmium selenide (CdSe) QDs via the high temperature decomposition method. Before the nucleation of the CdSe was allowed to take place, a solution of MNPs was injected into the solution, causing the QD to directly bind onto the MNP. The complex was monodispersed, crystalline, with an excitation of 575 nm and emission of 604 nm, and a quantum yield of 5%. The synthesis conditions were then varied in order to optimize the multifunctional nanomaterial produced. By varying the temperatures, injection rate and surfactant composition, created changes in the nanomaterials size, photoluminescence and morphology.

2.4 Experimental details

For this study magnetic-luminescent multifunctional nanocomposite material was synthesized. Following the work of Wang et al. [55], the QDs and MNPs were prepared separately. For the Fe_3O_4 MNP synthesis the co-precipitation method was chosen. The QDs were synthesized using the rapid hot injection method, we used InP\ZnSe because the study by Brunetti et al. [56] demonstrated that the In-based core-shell QDs are safer for *in vitro* and *in vivo* analysis than Cd-based QDs. The toxicity assessments found that the Cd-based QDs caused cell membrane damaged genetic material and interferes with Ca^{2+} homeostasis. The QDs were synthesized using the rapid hot injection technique and then a ligand exchange was performed on them and the resultant QDs were capped with 3-mercaptopropionic acid (MPA). The MNPs were functionalized with meso-2,3-dimercaptosuccinic acid (DMSA). This functionalization was achieved by creating a solution of 30 mM of DMSA in dimethyl sulfoxide (DMSO). This solution was added to a 40 mM of MNPs in toluene, at a 1:1 volume ratio. The resultant solution was stirred until it was observed that a black precipitate was forming. This black precipitate is the newly thiol-capped MNPs. These MNPs were removed from the solution with a magnet, washed with PBS, and dried in an oven. Using thiol chemistry, the QDs will be allowed to bond to the surface of the MNPs. Jeong et al. [57] was able to prepare multifunctional material using a similar synthesis method. The synthesis between the QDs and MNPs was accomplished through a partial ligand exchange reaction.

3. Characterization

The synthesized iron oxide magnetic-luminescent nanocomposite, was characterized using high resolution transmission electron microscopy (HR-TEM), photoluminescence (PL), and superconducting quantum interference device (SQUID) analysis.

The multifunctional nanocomposite material was synthesized using a partial ligand exchange. Using the partial ligand exchange method multifunction nanoclusters are formed; this occurred due to using InP\ZnSe in excess. The choice to use excess QDs was an attempt to reduce the quenching of fluoresce quantum dots caused by the MNPs. As seen by the HR-TEM image (**Figure 3**) the use of the InP/ZnSe in large excess compared to a number of MNPs led to enormous particle crowding. This crowding scenario makes it difficult to determine the average size of nanocomposite particles since the quantum dots filled the spaces between the MNPs.

The PL spectrum the MNP-QD nanocomposite confirms that the composite was successfully formed. In this study, it was discovered that in spite of the high ratio of the QDs to MNPs, the black MNPs quenched the fluorescing capability of the QDs, the lowered intensity is observed in the PL spectrum (**Figure 4**). The quenching could be possibly due to energy transfer process resulting from contact between the quantum dots and the surface of the iron oxide particles.

It is also clear that the absorption peak red shifted to 676 nm. This observation was also sufficient evidence for the successful formation of the nanocomposite, as the red shift emission observed in this study is most likely caused by the modification at the surface of the QDs brought by hydrophilic ligands and also immobilization of the MNPs (**Figure 5**).

The nanocomposite material maintained its magnetic properties after the MNPs were conjugated to the QDs as shown by, **Figure 6**, the magnetization curve. Saturation magnetization of Fe_3O_4-InP/ZnSe core-shell nanocomposite ~5.7 emu/g. After the MNP's were conjugated to the QD's the saturation magnetization is now a 10% fraction of the original MNPs.

The nanocomposites were exposed to both MCF-12A and KMST 6 cell lines. The cytotoxicity of the nanocomposite is presented in **Figure 7** below.

As shown by the **Figure 7** the cell viability was greater than 90% for all concentrations of the nanocomposite. The findings suggest that the nanocomposites are less toxic.

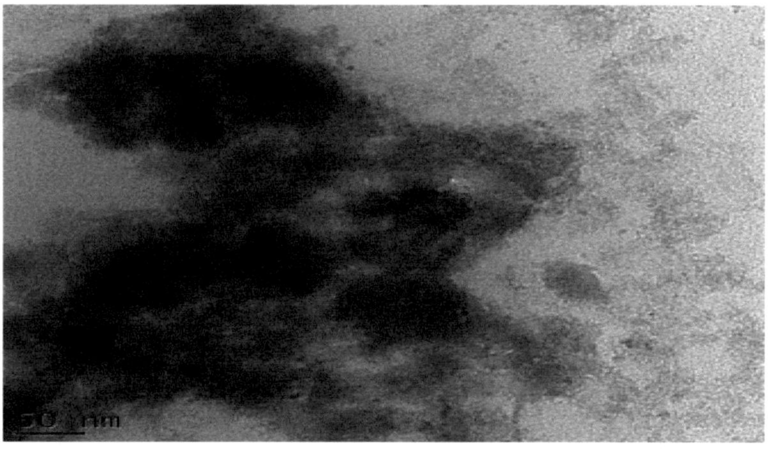

Figure 4.
MNPs-QDs nanocomposite.

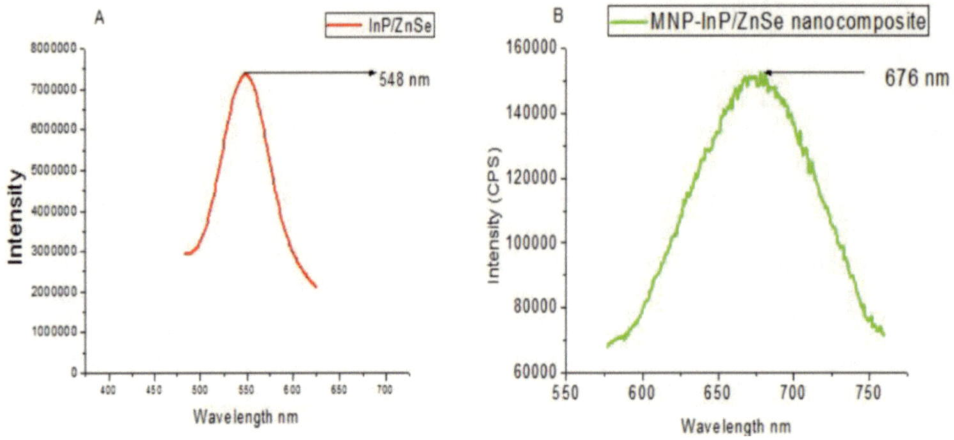

Figure 5.
PL spectra of InP/ZnSe nanocrystals dispersed in hexane (A) PL spectra of α-Fe_2O_3-InP/ZnSe nanocomposite dispersed in PBS (B) [58].

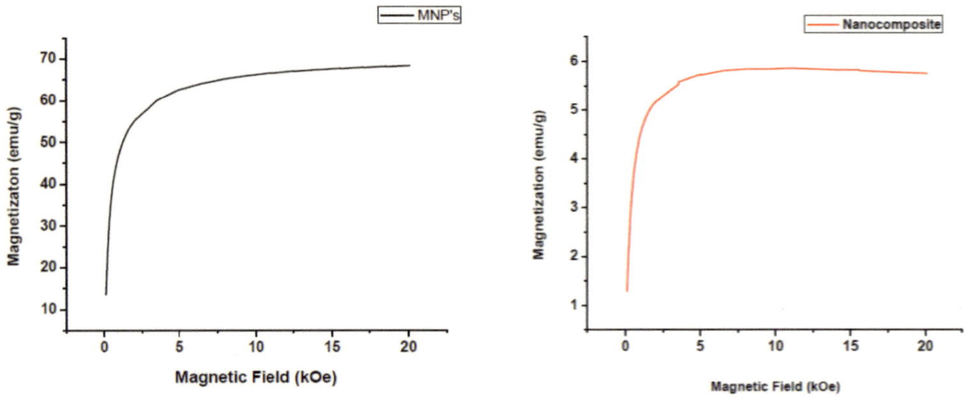

Figure 6.
Magnetization curves of the MNPs and the nanocomposite material.

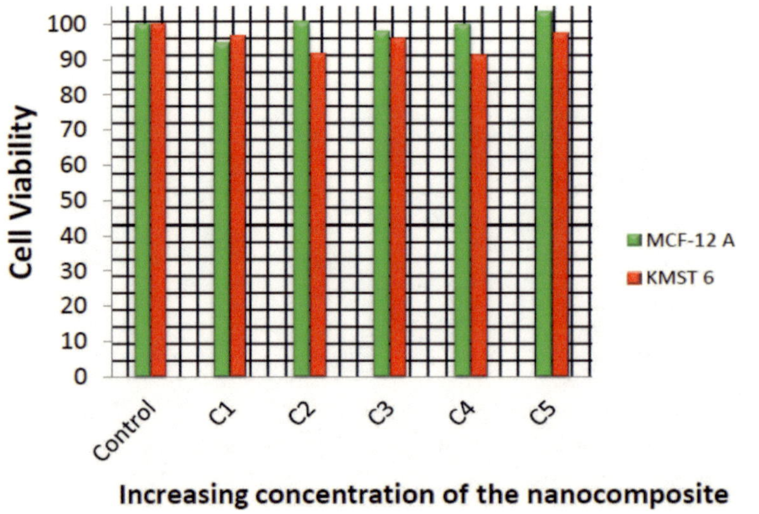

Figure 7.
Cytotoxicity of the iron oxide-InP/ZnSe nanocomposite [58].

4. Biomedical application of magnetic nanoparticles and fluorescent magnetic nanocomposites

Increasing attention has been drawn to the synthesis of MNPs for various applications. Magnetic nanomaterials have been observed to possess several distinctive characteristics, these unique capabilities have inspired many ideas in a wide range of biomedical applications [34, 58–68]. These applications include, [34, 63–71], target drug delivery [72], magnetic resonance imaging (MRI) contrasting agent [73, 74], cancer and tumor diagnosis and treatment [75]. Magnetic nanoparticles have demonstrated that they can be manipulated with an external magnetic field and thus to some extent be controlled to successfully reach a specific site of interest in a biological system. It has also been discovered that passing an alternating magnet field over magnetic nanomaterials causes them to heat up; this property makes them very attractive for therapies like hyperthermia, a treatment of cancer that requires selective heating to destroy cancer cells. This property also makes them promising for drug release treatment. Studies have also concluded that superparamagnetic nanoparticles can also improve magnetic resonance imaging (MRI) results. In MRI, aqueous dispersions of superparamagnetic IONPs have shown to be promising contrast agents, since it provides high-resolution images. This characteristic makes it possible to use IONPs as vector in a tracking device for gene and drug delivery. However, most methods require the use of superparamagnetic magnetite with particle size smaller than 20 nm [76]. Over recent years, MNPs have drawn a great deal of interest in cancer treatment, particularly IONPs. Studies have proved that IONPs can easily move into the cells with low cytotoxicity. They possess novel magnetic properties for drug delivery, cell targeting, imaging, tissue engineering and magnetofection. Cancer is known as one of the major causes of death worldwide and survival rates are still significantly low. Great research efforts have been devoted to improving the sensitivity and accuracy of diagnostic treatment for earlier detection and high efficiency, however treatment options not as effective. Recently explored magnetic-fluorescing nanoparticles can be used as simple, efficient and multifunctional diagnostic tool based on MRI [77]. The fluorescent NP emits at certain wavelength appropriate for visual imaging using fluorescence imaging microscopy. The multifunctional nanocomposite will simultaneously allow optical tracking as well as magnetic manipulation of biological processes [78]. Fluorescent-magnetic nanoparticles can be treated as bimodal probes useful for studies of the biological objects using both MRI and fluorescence detection. Bimodal imaging agents serving both for MRI and fluorescence imaging are of special interest. Therefore, we provide a brief introduction on the applications IONPs and fluorescent-IONPs in biomedicine, particularly as contrast agents for MRI diagnosis.

4.1 Applications of IONPs

IONPs possess unique physicochemical characteristics, as well as superparamagnetic with high surface area, non-toxicity, and biocompatibility [15]. IONPs have effectively been applied in various in biomedical applications [34, 58–68], since they can selectively target a specific biological unit by applying an external magnetic field. Iron oxide nanoparticles of the type, Fe_3O_4, have shown to be promising candidate as a contrast agent for magnetic resonance imaging. This is due to superparamagnetic or paramagnetism which creates an outer magnetic field around itself when exposed to an external magnetic field; this permits the increase of image resolution and decreases aggregation of particles due to fast dephasing of the spins through a so-called magnetic susceptibility effect. This enhances the signal intensity to help distinguish between healthy and unhealthy cells [79–81]. Studies have reported IONPs as promising MRI contrast agents for in vivo rat studies.

In of the studies, rats were anesthetized and subcutaneous injection containing 2.5 mg (Fe_3O_4)/kg body weight of Fe_3O_4 samples was given every 6 hours into the right hand of the animal. MRI scans taken after every 6 hours showed accumulation occurred on the lymph nodes, however none was noticed on the left-hand side. The study proved successful imaging of lymphatic system using iron oxide as a contrast agent [82]. At present, numerous studies are still undergoing clinical trials and only two types of dextran-capped IONPs have been clinically approved as MRI contrast agents, highlighted in **Table 3** [83]. These two are commonly known as, Ferucarbotran (Resovist) with particle size of about 60 nm, and Ferumoxides (Feridex in America and Endorem in Europe) have a broader particle size distribution between 120 and 180 nm (**Table 3**) [82, 83].

4.2 Application of fluorescent magnetic NPs

Over the years, scientists have shown that one way to improve on current nanomaterials was to combine two or more desired physical properties into one structure. The wish sparked many research ventures into the synthesis or assembly of these type multifunctional materials, also how many entities is effective and which areas could benefit most from these nanocomposites. Incorporation of a fluorescent material within a magnetic NP might modify its band gap energy as well as the luminescence properties [84]. Such multimodal properties are highly desirable specifically in the biomedical diagnosis and therapy [85, 86]. This nanocomposite would not only be improving current applications, but find better ways to achieve a desired outcome. These magnetic-fluorescent nanocomposites could be multimodal assays for in vitro- and in vivo-bioimaging applications such as MRI and fluorescence microscopy [27]. Other exciting applications of these nanocomposites include cell tracking, cytometry and magnetic separation, which could be easily controlled and monitored using fluorescent or confocal microscopy and molecular resonance imaging (MRI) [24, 87, 88]. They could also be used as bimodal agents for cancer therapy, additionally encompassing hyperthermic and photodynamic properties [89]. These fluorescent-magnetic nanocomposites can also be utilized as a multimodal therapeutic and diagnostic tool that can simultaneously locate, diagnose and treat various diseases [90–92]. In another study, Mandal et al. prepared multifunctional nanobiocomposite for targeted drug delivery in cancer therapy. Iron oxide nanoparticle of 15 nm in diameter was used as a contrast agent to enhance MRI and the anticancer drug gemcitabine. In vitro studies between treated and untreated cancer cell lines showed black spots on the gastric cancer cell lines that were treated with the nanobiocomposites whereas no reduction in the signal of the untreated cells. The study concluded that the iron oxide nanobiocomposite can act as contrast agent in MRI and also as a targeted drug delivery system in vivo using rats as an animal model [89]. In similar study, Ahmed et al. prepared the thiol capped-CdTe QDs coated with CTAB. The nanocomposites showed distinct magnetic and fluorescent properties even after isolation with a magnet it still maintained good PL intensity. The nanocomposite was conjugated to antibodies for the imaging of the colon carcinoma cells. No green fluorescence was observed on the surface of the cells. In vitro studies showed low toxicity at 64 fold dilutions. This demonstrated their potential as probes for imaging and ultimately provides a new class of multimodal diagnostics NPs for the complex biological systems [93, 94]. Hence, we focus on the developments of magnetic-fluorescent nanocomposites and their biological applications specifically, multimodal imaging for breast cancer diagnostics.

4.2.1 Multimodal bioimaging

Biological imaging or bioimaging is defined as the study of biological processes at the cellular and/or and subcellular level. Several biological imaging techniques

Names	Company	Applications	Relaxometric properties	Coating agent	Hydrodynamic size (nm)
Clariscan ferristene	GE Healthcare	Oral GI imaging	N/A	Sulfonated styrenedivinylbenzene copolymer	3500
Ferumoxsil AMI-121	Guerbet, advanced magnetics	Oral GI imaging	N/A	Silicon	300
Ferumoxides AMI-25	Guerbet, advanced magnetics	Liver imaging, cellular labeling	r1 = 10.1 r2 = 120	Dextran T10	120–180
Lumirem/Gastromark Ferucarbotran SHU-555A	Schering	Liver imaging, cellular labeling	r1 = 9.7 r2 = 189	Carboxydextran	60
Ferumoxytol code 7228	Advanced magnetics	Macrophage imaging blood pool agent, cellular labeling	r1 = 15 r2 = 89	Carboxylmethyldextran	30
Endorem/feridex ferumoxtran-10 AMI-227	Guerbet, advanced magnetics	Metastatic lymph node imaging	r1 = 9.9 r2 = 65	Dextran T10, T1	15–30
Resovist SHU-555C Supravist	Schering	Blood pool agent, cellular labeling	r1 = 10.7 r2 = 38	Carboxydextran	21
Feruglose NC100150	GE Healthcare (discontinue)	Blood pool agent	N/A	PEGylated starch	20
VSOP-C184	Ferropharm	Blood agent, cellular labeling	r1 = 14 r2 = 33.4	Citrate	7

Data edited from [82].

Table 3.
Characteristics of SIONPs agents undergoing clinical investigation clinical or commercial investigation.

have been developed with different in principles and equipment such as optical bioluminescence, optical fluorescence, ultrasound imaging, MRI, single-photon-emission computed tomography (SPECT), and positron emission tomography (PET), X-ray, thermal imaging, X-ray computed tomography (CT), hyperspectral imaging, and magnetic resonance imaging (MRI) [82]. Over three decades, these techniques have continuously had rapid developments and incremental improvements due to their wide application various biological fields. Multimodal magnetic nanoparticles have significant features as they could act as imaging probes and drug delivery systems. These NPs offer unique characteristic as a dual contrast agent that can combine fluorescent microscopy and MRI. Both techniques are well studied, MRI have been widely applied for in vivo imaging diagnosis, meanwhile fluorescence microscopy are mostly applied for in vitro imaging. Optical imaging is a promising tool as it provides better spatial resolution and performance in sensibility for in vitro imaging, however tissue penetration is limited to few millimeters. Moreover, MRI provides excellent spatial resolution and deep tissue contrast for better in vivo imaging. The amalgamation fluorescent microscopy and MRI opens new possibilities of rapid analysis for diagnosis of diseases and pathogens. In recent years, significant advances have been made in development of fluorescent magnetic nanoparticles as multimodal agents by using magnetic contrast agents. Zhang et al. prepared fluorescent mesoporous silica coated-iron oxide nanoparticles of ~10 nm with high magnetic resonance sensitivity and excellent cell labeling efficiency for detection of neural progenitor cells using MRI [95, 96]. In another study, monodispersed magnetic nanoparticles functionalized with an organic dye showed optical activity and good biocompatibility [93, 94]. In recent study, Guo et al. synthesized superparamagnetic monodispersed core@shell $CoFe_2O_4$@$MnFe_2O_4$ NPs coated with poly(isobutylene-alt-maleic anhydride) PEG and then functionalized with folic acid. The resulting multifunctional

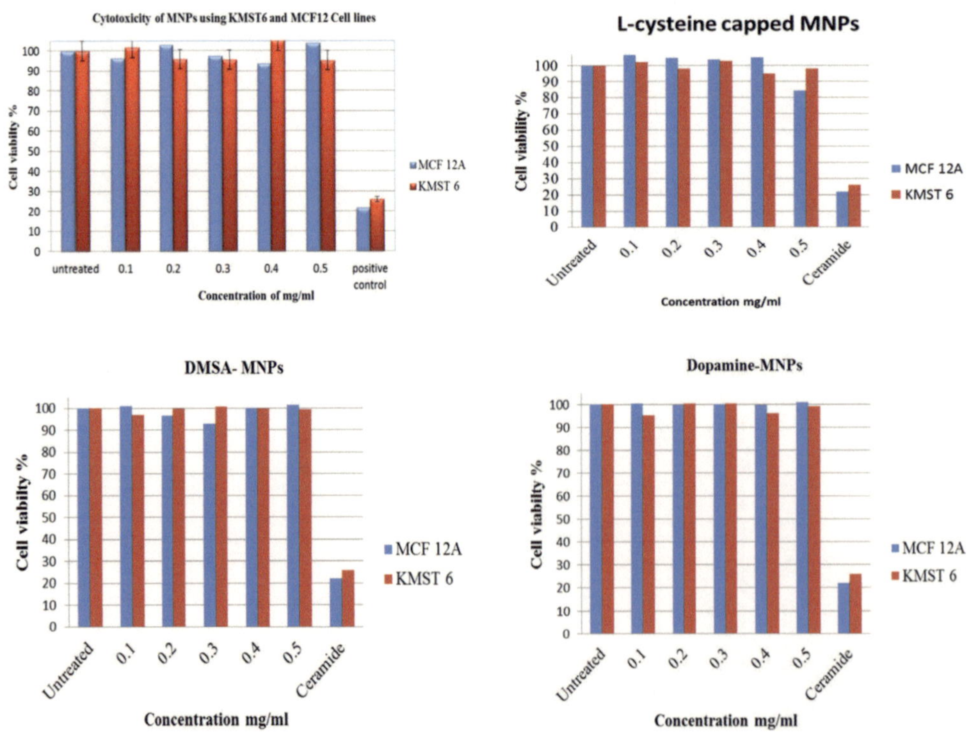

Figure 8.
Cytotoxicity studies of bare and functionalized iron oxide nanoparticles using MCF-12A and KMST 6 cell lines.

nanocomposite exhibit good biocompatibility, high T_2 relaxation, and long-term fluorescence stability to enhance the targeted MRI and fluorescent tracking for in vivo and in vitro studies [97, 98]. Recent research advancements have produced several excellent magnetic fluorescent nanocomposites. In our study, we prove genes are capable of being used for potential development of serum markers in the diagnosis, of breast cancer or early detection of poor-outcome breast cancer. However, we found that these proteins are present in a very low concentration, which makes the diagnosis a challenging process, however not impossible. The synthesized method used for the preparation of bare and capped iron oxide nanoparticles show low cytotoxicity, which is a strong foundation for future contrast agent for in vivo studies of certain cancer therapies. In addition, it has been proved that different ligand molecules have different effects on the toxicity of the nanoparticles, for example, Liu et al. [93, 94] carried out cytotoxicity study of iron oxide coated with acridine orange using 293 T cells. They varied the concentration of the iron oxide from 0 to 80 mg/mL and observed cell viability greater than 78%. In similar example, In comparison. The lower cell viability observed here could be attributed to acridine orange used and not the iron oxide. In a very recent study, Zhang et al. synthesized superparamagnetic Fe_2O_3 NPs with a diameter of 51.88 nm showed neurotoxic effects in PC12 cell line, in a dose-dependent manner at 60–200 mg/mL, but not at 10–50 mg/mL [99]. The ligands chosen in our study appeared not to affect the toxicity of the iron oxide nanoparticles despite using higher concentration compared to the concentration reported by Liu et al. and Zhang et al. (**Figure 8**) [93, 94, 99].

5. Conclusion

We were able to successfully synthesize iron oxide magnetic nanoparticles using the co-precipitation method. The synthesized nanoparticles were then functionalized with DMSA.

We were also able to successfully synthesize InP/ZnSe nanocrystals using the hot injection method. The synthesized nanocrystals were capped with oleic acid, which was the stabilizing agent in the nanocrystals' synthesis. The InP/ZnSe then underwent a ligand exchange and thus the oleic acid capped QDs were replaced with MPA capped QDs.

The ultimate objective of the study was realized when we successfully fabricated a magnetic-luminescent bifunctional nanocomposite material was prepared using thiol-chemistry, this allowed the direct combination of the QDs and MNPs. The nanocomposite material was characterized and observed to exhibit both magnetic and luminescent properties. The SQUID analysis showed that the Fe_3O_4-InP/ZnSe nanocomposite material has a magnetic saturation of 6.03 emu/g. The PL studies demonstrated that the nanocomposite material had a fluorescence of approximately 40,000 arbitrary units. The nanocomposite material had significantly lower magnetic and fluorescence properties in comparison to their pure forms.

To conclude the study we carried out extensive in vitro cytotoxicity study to evaluate the toxicity of the iron oxide nanoparticles, functionalized iron oxide nanoparticles, InP/ZnSe nanocrystals, and Fe_2O_3-InP/ZnSe nanocomposite. The KMST 6 and MCF-12A cell lines were exposed to increasing concentration of the nanoparticles. The cells were incubated with the nanoparticles for 24 hours and the cell viability was determined using MTT assay. The cell viability for all types of the nanomaterials was greater than 90% using both MCF-12A and KMST6 cell lines. This suggested that the particles are safe hence not limiting their biological applications and also safe in regard to handling.

Author details

Martin Onani*, Leandre Brandt and Zuraan Paulsen
Department of Chemistry, University of the Western Cape, Bellville, South Africa

*Address all correspondence to: monani@uwc.ac.za

References

[1] Corot C, Robert P, Idee JM, Port M. Recent advances in iron oxide nanocrystal technology for medical imaging. Advanced Drug Delivery Reviews. 2006;**58**:1471-1504

[2] Sun C, Lee JSH, Zhang M. Magnetic nanoparticles in MR imaging and drug delivery. Advanced Drug Delivery Reviews. 2008;**60**:1252-1265

[3] Veiseh O, Gunn JW, Zhang M. Design and fabrication of magnetic nanoparticles for targeted drug delivery and imaging. Advanced Drug Delivery Reviews. 2010;**62**:284-304

[4] Lee JH, Lee K, Moon SH, Lee YH, Park TG, Cheon J. All-in-one target-cell-specific magnetic nanoparticles for simultaneous molecular imaging and siRNA delivery. Angewandte Chemie. 2009;**48**:4174-4179

[5] Tsai ZT, Wang JF, Kuo HY, Shen CR, Wang JJ, Yen TC. In situ preparation of high relaxivity iron oxide nanoparticles by coating with chitosan: A potential MRI contrast agent useful for cell tracking. Journal of Magnetism and Magnetic Materials. 2010;**322**:208-213

[6] Safarık I, Safarıkova M. Magnetic techniques for the isolation and purification of proteins and peptides. BioMagnetic Research and Technology. 2004;**2**:7-34

[7] Zhao Z, Bian Z, Chen L, He X, Wang Y. Synthesis and surface-modifications of iron oxide magnetic nanoparticles and applications on separation and analysis. Progress in Chemistry. 2006;**18**:1288-1297

[8] Ito A, Tanaka K, Honda H, Abe S, Yamaguchi H, Kobayashi T. Complete regression of mouse mammary carcinoma with a size greater than 15 mm by frequent repeated hyperthermia using magnetite nanoparticles. Journal of Bioscience and Bioengineering. 2003;**96**:364-369

[9] Salloum M, Ma R, Zhu L. Enhancement in treatment planning for magnetic nanoparticle hyperthermia: Optimization of the heat absorption pattern. International Journal of Hyperthermia. 2009;**25**:309-321

[10] Kaushik A, Solanki PR, Ansari AA, Malhotra BD, Ahmad S. Iron oxide-chitosan hybrid nanobiocomposite based nucleic acid sensor for pyrethroid detection. Biochemical Engineering Journal. 2009;**46**:132-140

[11] Lacroix LM, Ho D, Sun SH. Magnetic nanoparticles as bothimaging probes and therapeutic agents. Current Topics in Medicinal Chemistry. 2010;**10**(12):1184-1197

[12] Lamanna G, Kueny-Stotz M, Mamlouk-Chaouachi H, Ghobril C, Basly B, Bertin A, et al. Dendronized iron oxide nanoparticles for multimodal imaging. Biomaterials. 2011;**32**(33):8562-8573

[13] Kim BH, Lee N, Kim H, An K, Park YI, Choi Y, et al. Large-scale synthesis of uniform and extremely small-sized iron oxide nanoparticles for high-resolution T-1 magnetic resonance imaging contrast agents. Journal of the American Chemical Society. 2011;**133**(32):12624-12631

[14] Ling D, Lee N, Hyeon T. Chemical synthesis and assembly of uniformly sized iron oxide nanoparticles for medical applications. Accounts of Chemical Research. 2015;**48**(5):1276-1285

[15] Sugimoto T. Formation of monodispersed nano- and micro-particles controlled in size, shape, and internal structure. Chemical Engineering and Technology. 2003;**26**(3):313-321

[16] Colombo M, Carregal-Romero S, Casula MF, Gutierrez L, Morales MP, Bohm IB, et al. Biological applications of magnetic nanoparticles. Chemical Society Reviews. 2012;**41**(11):4306-4334

[17] Massart R. Preparation of aqueous magnetic liquids in alkaline and acidic media. IEEE Transactions on Magnetics. 1981;**17**(2):1247-1248

[18] Massart R, Dubois E, Cabuil V, Hasmonay E. Preparation and properties of monodisperse magnetic fluids. Journal of Magnetism and Magnetic Materials. 1995;**149**(1-2):1-5

[19] Hui C, Shen CM, Yang TZ, Bao LH, Tian JF, Ding H, et al. Large-scale Fe_3O_4 nanoparticles soluble in water synthesized by a facile method. Journal of Physical Chemistry C. 2008;**112**(30):11336-11339

[20] Chin AB, Yaacob II. Synthesis and characterization of magnetic iron oxide nanoparticles via w/o microemulsion and Massart's procedure. Journal of Materials Processing Technology. 2007;**191**(1-3):235-237

[21] Cai W, Wan JQ. Facile synthesis of superparamagnetic magnetite nanoparticles in liquid polyols. Journal of Colloid and Interface Science. 2007;**305**(2):366-370

[22] Xia T, Wang JP, Wu CL, Meng FC, Shi Z, Lian J, et al. Novel complex-coprecipitation route to form high quality triethanolamine-coated Fe_3O_4 nanocrystals: Their high saturation magnetizations and excellent water treatment properties. CrystEngComm. 2012;**14**(18):5741-5744

[23] Xing R, Liu G, Zhu J, Hou Y, Chen X. Functional magnetic nanoparticles for non-viral gene delivery and MR imaging. Pharmaceutical Research. 2014;**31**(6):1377-1389

[24] Corr SA, Rakovich YP, Gun'ko YK. Multifunctional magnetic-fluorescent nanocomposites for biomedical applications. Nanoscale Research Letters. 2008;**3**:87-104

[25] Xu Y, Karmakar A, Wang D, Mahmood MW, Watanabe F, Zhang Y, et al. Multifunctional Fe_3O_4 cored magnetic-quantum dot fluorescent nanocomposites for RF nanohyperthermia of cancer cells. Journal of Physical Chemistry C. 2010;**114**:5020-5026

[26] Lim E-K, Yang J, Dinney CPN, Suh J-S, Huh Y-M, Haam S. Self-assembled fluorescent magnetic nanoprobes for multimode-biomedical imaging. Biomaterials. 2010;**31**:9310-9319

[27] Laurent S, Forge D, Port M, Roch A, Robic C, Vander Elst L, et al. Chemical Reviews. 2008;**108**:2064-2110

[28] Ali K, Javed Y, Jamil Y. Size and shape control synthesis of ion oxide-based nanoparticles: Current status and future possibility. In: Sharma SK, editor. Complexnmagnetic Nanostructures. Cham: Springer; 2017. pp. 39-81

[29] Ali A, Zafar H, Zia M, Ul Haq I, Phull AR, Ali JS, et al. Synthesis, characterization, applications, and challenges of iron oxide nanoparticles. Nanotechnology, Science and Applications. 2016;**9**:49-67

[30] Campos EA, Pinto DVBS, de Oliveira JIS, Mattos ED, Dutra RDL. Synthesis, characterization and applications of iron oxide nanoparticles—A short review. Journal of Aerospace Technology and Management. 2015;7:267-276

[31] Wu W, Wu Z, Yu T, Jiang C, Kim WS. Recent progress on magnetic iron oxidenanoparticles: Synthesis, surface functional strategies and biomedical applications. Science and Technology of Advanced Materials. 2015;**16**:023501

[32] Tartaj P, Morales MP, González-Carreño T, Veintemillas-Verdaguer S, Serna CJ. Advances in magnetic nanoparticles for biotechnology applications. Journal of Magnetism and Magnetic Materials. 2005;**290-291**:28-34

[33] Sapsford KE, Algar WR, Berti L, Gemmill KB, Casey BJ, Oh E, et al. Functionalizing nanoparticles with biological molecules: Developing chemistries that facilitate nanotechnology. Chemical Reviews. 2013;**113**:1904-2074

[34] Gupta AK, Gupta M. Synthesis and surface engineering of iron oxide nanoparticles for biomedical applications. Biomaterials. 2005;**26**:3995-4021

[35] Bandhu A, Mukherjee S, Acharya S, Modak S, Brahma SK, Das D, et al. Dynamic magnetic behavior and Mössbauer effect measurements of magnetite nanoparticles prepared by a new technique in the co-precipitation method. Solid State Communications. 2009;**149**:1790-1794

[36] Ozkaya T, Toprak MS, Baykal A, Kavas H, Koseoglu Y, Aktas B. Synthesis of Fe_3O_4 nanoparticles at 100°C and its magnetic characterization. Journal of Alloys and Compounds. 2009;**472**:18-23

[37] Zhao Z, Zhou Z, Bao J, Wang Z, Hu J, Chi X, et al. Octapod iron oxide nanoparticles as high-performance T_2 contrast agents for magnetic resonance imaging. Nature Communications. 2013;**4**:2266. DOI: 10.1038/ncomms3266

[38] Song L, Zhang W, Chen H, Zhang X, Wu H, Ma M, et al. International Journal of Nanomedicine. 2019;**14**:921-936

[39] Guoa AL, Chenc H, Hea N, Denga Y. Effects of surface modifications on the physicochemical properties of iron oxide nanoparticles and their performance as anticancer drug carriers. Chinese Chemical Letters. 2018;**29**(12):1829-1833. DOI: 10.1016/j.cclet.2018.10.038

[40] Kumar B, Smita K, Cumbal L, Debut A, Galeas S, Guerrero VH. Phytosynthesis and photocatalytic activity of magnetite (Fe_3O_4) nanoparticles using the Andean blackberry leaf. Materials Chemistry and Physics. 2016;**179**:310-315

[41] Kiplagat A, Onani MO, Meyer M, Akenga TA, Dejene FB. Synthesis and characterization of luminescence magnetic nanocomposite. Physica B: Condensed Matter. 2015:1-9. DOI: 10.1016/j.physb.2015.08.037

[42] Monaco I, Arena F, Biffi S, Locatelli E, Bortot B, La Cava F, et al. Synthesis of lipophilic core-shell Fe_3O_4@SiO_2@Au nanoparticles and polymeric entrapment into nanomicelles: A novel nanosystem for in vivo active targeting and magnetic resonance-photoacoustic dual imaging. Bioconjugate Chemistry. 2017;**28**(5):1382-1390. DOI: 10.1021/acs.bioconjchem.7b00076

[43] Toropova YG, Golovkin AS, Malashicheva AB, Korolev DV, Gorshkov AN, Gareev KG, et al. In vitro toxicity of Fe_mO_n, FemOn-SiO_2 composite, and SiO_2-FemOn core-shell magnetic nanoparticles. International Journal of Nanomedicine. 2017;**12**:593-603. DOI: 10.2147/IJN.S122580

[44] Kim DK, Mikhaylova M, Zhang Y, Muhammed M. Protective coating of superparamagnetic iron oxide nanoparticles. Chemistry of Materials. 2003;**15**:1617-1627

[45] Xu S, Kumar S, Nann T. Rapid synthesis of high-quality InP nanocrystals. American Chemical Society. 2006;**128**:1054-1055

[46] Khanna PK, Juna K, Honga KB, Baega J, Mehrotra GK. Synthesis of indium phosphide nanoparticles via

catalytic cleavage of phosphorus carbon bond in n-trioctylphosphine by indium. Materials Chemistry and Physics. 2005;**92**:54-58

[47] Lui B, Huang W, Wang D, Xie W, Yu M, Yao A. Fabrication of luminescent-superparamagnetic CdSe-QDs/SiO_2/ Fe_3O_4 nanocomposite particles. NSIT-Nanotech. 2008;**2**:297-301

[48] Liu D, Tong L, Shi J, Yang H. Luminescent and magnetic properties of YVO4:Ln^{3+}@Fe_3O_4 (Ln^{3+}=Eu^{3+} or $Dy^{3+)}$ nanocomposites. Journal of Alloys and Compounds. 2012;**512**:361-365

[49] Liu Z, Kumbhar A, Xu D, Zhang J, Sun Z, Fang J. Coreduction colloidal synthesis of III–V nanocrystals: The case of InP. Angewandte Chemie International. 2008;**47**:3540-3542

[50] Hong X, Li J, Wang M, Xu J, Guo W, Li J, et al. Fabrication of magnetic luminescent nanocomposites by a layer-by-layer self-assembly approach. Chemistry of Materials. 2004;**16**:4022-4027

[51] Gua J, Yang W, Wang C, He J, Chen J. Poly(N-isopropylacrylamide)-coated luminescent/magnetic silica microspheres: Preparation, characterization, and biomedical applications. Chemistry of Materials. 2006;**18**:5554-5562

[52] Nai-Qiang Y, Ling L, Jie-Mei L, Yan-Song L, Mao-Gang G, Yi-Zhi W, et al. Preparation and characterization of nontoxic magnetic-luminescent nanoprobe. Chinese Physics B. 2012;**21**(11):116101-116106

[53] Zedan AF, Abdelsayed V, Mohamed MB, Samy E-SM. Rapid synthesis of magnetic/luminescent (Fe_3O_4/CdSe) nanocomposites by microwave irradiation. Journal of Nanoparticle Research. 2013;**15**:1312

[54] ChoM,ContrerasEQ,LeeSS,JonesCJ, Jang W, Colvin VL. Characterization and optimization of the fluorescence of nanoscale iron oxide/quantum dot complexes. Journal of Physical Chemistry C. 2014;**118**:14606-14616

[55] Wang D, He J, Rosenzweig N, Rosenzweig Z. Superparamagnetic Fe_2O_3 beads-CdSe/ZnS quantum dots core-shell nanocomposite particles for cell separation. Nano Letters. 2004;**4**(3):409-413

[56] Brunetti V, Chibli H, Fiammengo R, Galeone A, Malvindi MA, Vecchio G, et al. InP/ZnS as a safer alternative to CdSe/ZnS core/shell quantum dots: In vitro and in vivo toxicity assessment. Nanoscale. 2012:1-11

[57] Jeong J, Kwon E, Cheong T, Park H, Cho N, Kim W. Synthesis of multifunctional Fe_3O_4–CdSe/ZnS nanoclusters coated with lipid a toward dendritic cell-based immunotherapy. ACS Applied Materials & Interfaces. 2014;**6**:5297-5307

[58] Ayabei K. Fabrication of luminescent magnetic nanocomposite for diagnosis of breast cancer [A thesis submitted to the Faculty of Natural Sciences]. University of the Western Cape; 2016

[59] Hilger I, Hergt R, Kaiser WA. Use of magnetic nanoparticle heating in the treatment of breast cancer. IEE Proceedings Nanobiotechnology. 2005;**152**:33-39

[60] Du K, Zhu YH, Xu HB, Yang XL. Multifunctional magnetic nanoparticles: Synthesis, modification and biomedical applications. Progress in Chemistry. 2011;**23**:2287-2298

[61] Huang SH, Juang RS. Biochemical and biomedical applications of multifunctional magnetic nanoparticles: A review. Journal of Nanoparticle Research. 2011;**13**:4411-4430

[62] Schladt TD, Schneider K, Schild H, Tremel W. Synthesis and bio-functionalization of magnetic nanoparticles for medical diagnosis and treatment. Dalton Transactions. 2011;**40**:6315-6343

[63] Mahmoudi M, Serpooshan V, Laurent S. Engineered nanoparticles for biomolecular imaging. Nanoscale. 2011;**3**:3007-3026

[64] Le Trequesser Q, Seznec H, Delville MH. Functionalized nanomaterials: Their use as contrast agents in bioimaging: Mono- and multimodal approaches. Nanotechnology Reviews. 2013;**2**:125-169

[65] Yan K, Li PH, Zhu HE, Zhou YJ, Ding JD, Shen J, et al. Recent advances in multifunctional magnetic nanoparticles and applications to biomedical diagnosis and treatment. RSC Advances. 2013;**3**:10598-10618

[66] Padmanabhan P, Kumar A, Kumar S, Chaudhary RK, Gulyas B. Nanoparticles in practice for molecular-imaging applications: An overview. Acta Biomaterialia. 2016;**41**:1-16

[67] Cherukula K, Lekshmi KM, Uthaman S, Cho K, Cho CS, Park IK. Multifunctional inorganic nanoparticles: Recent progress in thermal therapy and imaging. Nanomaterials. 2016;**6**:76

[68] Nishio K, Ikeda M, Gokon N, Tsubouchi S, Narimatsu H, Mochizuki Y, et al. Preparation of size-controlled (30-100 nm) magnetite nanoparticles for biomedical applications. Journal of Magnetism and Magnetic Materials. 2007;**310**:2408-2410

[69] Jun YW, Lee JH, Cheon J. Chemical design of nanoparticle probes for high-performance magnetic resonance imaging. Angewandte Chemie, International Edition. 2008;**47**:5122-5135

[70] Sudimack JBA, Lee RJ. Targeted drug delivery via the folate receptor. Advanced Drug Delivery Reviews. 2000;**41**:147-162

[71] Hu F, Wei L, Zhou Z, Ran Y, Li Z, Gao M. Preparation of biocompatible magnetite nanocrystals for in vivo magnetic resonance detection of cancer. Advanced Materials. 2006;**18**:2553-2556

[72] Song H, Choi J, Huh Y, Kim S, Jun Y, Suh J, et al. Surface modulation of magnetic nanocrystals in the development of highly efficient magnetic resonance probes for intracellular labeling. Journal of the American Chemical Society. 2005;**127**:9992-9993

[73] Jordan A, Scholz R, Maier-hauff K, Johannsen M, Wust P, Nadobny J, et al. Presentation of a new magnetic field therapy system for the treatment of human solid tumors with magnetic fluid hyperthermia. Journal of Magnetism and Magnetic Materials. 2001;**22**:5118-5126

[74] Sun S, Zeng H. Size-controlled synthesis of magnetite nano-particles. Journal of the American Chemical Society. 2002;**124**:8204-8205

[75] Kalber TL, Smith CJ, Howe FA, Griffiths JR, Ryan AJ, Waterton JC, et al. Investigative Radiology. 2005;**40**: 784-791

[76] Iliuk AB, Hu L, Tao WA. Analytical Chemistry. 2011;**83**:4440-4452

[77] Mascini M, Palchetti I, Tombelli S. Angewandte Chemie (International Ed. in English). 2012;**51**:1316-1332

[78] Zhang B, Chen B, Wang Y, Guo F, Li Z, Shi D. Journal of Colloid and Interface Science. 2011;**353**: 426-432

[79] Dunn JF, Roche MA, Springett R, et al. Monitoring angiogenesis in brain using steady-state quantification of ΔR2 with MION infusion. Magnetic Resonance in Medicine. 2004;**51**(1):55-61

[80] Mandal A, Sekar S, Kanagavel M, Chandrasekaran N, Mukherjee A, Sastry TP. Biochimica et Biophysica Acta. 2013;**1830**:4628-4633

[81] Wang YXJ, Hussain SM, Krestin GP. Superparamagnetic iron oxide contrast agents: Physicochemical characteristics and applications in MR imaging. European Radiology. 2001;**11**(11):2319-2331

[82] Jing LH, Ding K, Kershaw SV, Kempson IM, Rogach AL, Gao MY. Magnetically engineered semiconductor quantum dots as multimodal imaging probes. Advanced Materials. 2014;**26**:6367-6386

[83] Abdullah Mirzaie R, Kamrani F, Anaraki Firooz A, Khodadadi AA. Materials Chemistry and Physics. 2012;**133**:311-316

[84] Kim J, Lee JE, Lee SH, Yu JH, Lee JH, Park TG, et al. Advanced Materials. 2008;**20**:478-483

[85] Lee JE, Lee N, Kim T, Kim J, Hyeon T. Accounts of Chemical Research. 2011;**44**:893-902

[86] Jana NR. Design and development of quantum dots and other nanoparticles based cellular imaging probe. Physical Chemistry Chemical Physics. 2011;**13**:385-396

[87] Koole R, Mulder WJM, van Schooneveld MM, Strijkers GJ, Meijerink A, Nicolay K. Magnetic quantum dots for multimodal imaging. Wiley Interdisciplinary Reviews. Nanomedicine and Nanobiotechnology. 2009;**1**:475-491

[88] Wang Y-XJ. Superparamagnetic iron oxide based MRI contrast agents: Current status of clinical application. Quantitative Imaging in Medicine and Surgery. 2011;**1**(1):35-40

[89] Gao JH, Gu HW, Xu B. Multifunctional magnetic nanoparticles: Design, synthesis, and biomedical applications. Accounts of Chemical Research. 2009;**42**:1097-1107

[90] Shi DL, Sadat ME, Dunn AW, Mast DB. Photo-fluorescent and magnetic properties of iron oxide nanoparticles for biomedical applications. Nanoscale. 2015;**7**:8209-8232

[91] Bronzino JD, Peterson DR. Biomedical Signals, Imaging, and Informatics. Boca Raton: CRC Press, Taylor & Francis Group; 2015

[92] Ahmed S, Dong J, Yui M, Kato T, Lee J, Park EY. Journal of Nanobiotechnology. 2013;**11**:28

[93] Carretta P, Lascialfari A. NMR-MRI, mSR and Mössbauer Spectroscopies in Molecular Magnets. Springer-Verlag Mailand: Milan, Italy; 2007

[94] Liu CH, Sahoo SL, Tsao MH. Colloids and Surfaces. B, Biointerfaces. 2014;**115**:150-156

[95] Acharya A. Luminescent magnetic quantum dots for In vitro/ In vivo imaging and applications in therapeutics. Journal of Nanoscience and Nanotechnology. 2013;**13**:3753-3768

[96] Zhang L, Wang Y, Tang Y, Jiao Z, Xie C, Zhang H, et al. High MRI performance fluorescent mesoporous silica-coated magnetic nanoparticles for tracking neural progenitor cells in an ischemic mouse model. Nanoscale. 2013;**5**:4506-4516

[97] Catherine CB, Adam SGC. Functionalisation of magnetic

nanoparticles for applications in biomedicine. Journal of Physics D: Applied Physics. 2003;**36**:R198

[98] Gao G, Zhang Q, Yin T, Shapter JG, Lai W, Huang P, et al. ACS Applied Materials & Interfaces. 2017;**9**:17777-17785

[99] Lu X, Jiang R, Fan Q, Zhang L, Zhang H, Yang M, et al. Fluorescent-magnetic poly(poly(ethyleneglycol) monomethacrylate)-grafted Fe_3O_4 nanoparticles from post-atom-transfer-radical-polymerization modification: Synthesis, characterization, cellular uptake and imaging. Journal of Materials Chemistry. 2012;**22**:6965-6973